Resources for Teaching

Convergences

message • method • medium

Ben McCorkle

BEDFORD/ST. MARTIN'S
Boston • New York

Manufactured in the United States of America.

6 5 4 3 2
f e d c b a

For information, write: Bedford/St. Martin's, 75 Arlington Street, Boston, MA 02116
(617-399-4000)

ISBN: 0–312–25073–8

Preface

Convergences is a new kind of reader, designed specifically for a new generation of students. It contains an extensive (perhaps unprecedented) sampling of visual and verbal material representing a wide variety of new and traditional media. This diverse material was collected for a practical instructional purpose: to equip students with the critical and creative tools they need to become better readers and writers within our rapidly changing cultural environment.

Over the past decade or so, we have witnessed a series of remarkable technological changes. These changes have affected nearly every aspect of social and cultural life, from how we communicate with others to how we perceive the world, from how we obtain information to how we buy merchandise. The changes have been quickly assimilated into the culture. Products and technologies that were largely unimaginable thirty or even twenty years ago—cell phones, the Internet, powerful desktop and handheld computers, compact disc players, mP3 players, video games, digital cameras—are now so familiar we barely notice their impact.

The convergence of the computer, the Internet, and new telecommunications systems, along with the development of high-speed, digital audiovisual media, has profoundly changed the ways in which we obtain and process information. High-tech professionals and computer experts have called attention to this convergence for some time, plugging "convergence products" such as the combination cell phone/handheld computer and promoting a future characterized by a "seamless, universal connectivity." But convergence is no longer confined to the world of computer technology. Its ramifications are being felt throughout the culture—in art, literature, education, entertainment, and media. This book explores the impact of all these convergences on our traditional notions of literacy. What are the prospects for the printed word in a culture that depends increasingly on visual media? As traditional print culture begins to blend into an emerging visual culture, how do we adjust to and interpret the new texts?

An expansion of literacy means acquiring a new understanding of what a "text" is and how it is "read." Our culture still retains a literary notion of text that usually privileges word over image. Images and graphic design are often considered merely as illustrations or visual embellishments that accompany a text, and not as integral elements of that text. Our critical vocabulary—the terms we use to interpret and evaluate the various forms of communications we experience daily—needs to be attuned to the flourishing visual culture in which we now live. Developing a critical awareness of new texts such as Web pages, electronic mail, multimedia encyclopedias, or interactive video games entails an understanding of how layout, design, imagery, sound, and animation interrelate with language to create messages that then depend on our ability to assimilate a total effect. It means reading and writing in a multimedia and multitasking environment, where print is only one textual tool among many for producing meaning. How do we read advertising texts that deliberately disguise themselves as editorials? Or commercials that pretend to be video games? What about essays that take cinematic or mosaic forms as their models or are juxtaposed with photographs? When an artwork incorporates words, do we "read" it in the same way we read a printed text?

The printed word, of course, remains and will continue to remain very much part of our lives. Many texts still consist almost entirely of continuous print. A best-selling novel will most likely be composed of several hundred pages of words printed in the same format and typeface. But other major print media such as newspapers and magazines have grown less print-oriented and more dependent on layout and design that integrate photographs, illustrations, advertising, and other graphic elements. Sometimes, especially in magazines targeted to young or trendy audiences, we find the printed word engulfed by design elements, making it difficult to disentangle the written article from the layout and design. The printed word can hardly be said to be endangered, but the contexts and environments in which it now appears have dramatically changed. The printed word now finds itself in many new situations, where it shares (or at times competes for) space with other media elements—pictures, icons, sound, graphic design, display, animation, and so on. It is not unusual to watch a news program in which a live interview is presented on a partitioned screen where other images and messages—some unrelated to the interview—are flashing by. The audience is expected to watch, listen, and read simultaneously.

Windowed multimedia, only one manifestation of the continuing transformation of our communication and informational networks, does not signal the death of reading and writing. On the contrary, these skills appear to be more important now than ever before. As readers and writers, students will be increasingly expected to handle new textual configurations, especially those that depend on a complex interplay of print, imagery, and design.

BIBLIOGRAPHY OF RELATED SOURCES

In addition to the many sources mentioned in the text, I found the following books and articles especially helpful:

Bachelard, Gaston. *The Poetics of Space*. Translated by Maria Jolas. 1953. Reprint, Boston: Beacon Press, 1994.

Barrett, Edward, and Marie Redmond, eds. *Contextual Media: Multimedia and Interpretation*. Cambridge, Mass.: MIT Press, 1995.

Barthes, Roland. *Image Music Text*. Translated by Stephen Heath. New York: Hill & Wang, 1977.

———. *The Semiotic Challenge*. Translated by Richard Howard. New York: Hill & Wang, 1988.

Baudrillard, Jean. *Selected Writings*. Edited by Mark Poster. Stanford, Calif.: Stanford University Press, 1988.

Bolter, Jay David, and Richard Grusin. *Remediation: Understanding New Media*. Cambridge, Mass.: MIT Press, 1999.

Calvino, Italo. *Six Memos for the Next Millennium*. Translated by Patrick Creagh. 1988. Reprint, New York: Vintage, 1996.

Corbett, Edward P. J. *Classical Rhetoric for the Modern Student*, 2nd ed. Oxford, England: Oxford University Press, 1971.

Danto, Arthur C. *Beyond the Brillo Box: The Visual Arts in Post-Historical Perspective*. New York: Farrar, Straus and Giroux, 1992.

Davis, Mike. *City of Quartz*. London: Verso, 1990.

Duany, Andres, Elizabeth Plater-Zyberk, and Jeff Speck. *Suburban Nation: The Rise of Sprawl and the Decline of the American Dream*. New York: North Point Press, 2000.

Earle, Richard. *The Art of Cause Marketing*. Lincolnwood, Ill.: NTC Business Books, 2000.

Ferguson, Niall, ed. *Virtual History: Alternatives and Counterfactuals*. New York: Basic Books, 1999.

Genette, Gerard. *Narrative Discourse Revisited*. Translated by Jane E. Lewin. Ithaca, N.Y.: Cornell University Press, 1988.

Hobsbawm, Eric. *The Age of Extremes*. New York: Pantheon, 1995.

Jameson, Fredric. *Postmodernism: Or, The Cultural Logic of Late Capitalism*. Durham, N.C.: Duke University Press, 1991.

Lanham, Richard A. *The Electronic Word: Democracy, Technology, and the Arts*. Chicago: University of Chicago Press, 1993.

Lutticken, Sven. "Art in the Age of Convergence." *New Left Review* 6 (November/December 2000).

MacRae-Gibson, Gavin. *The Secret Life of Buildings: An American Mythology for Modern Architecture*. Cambridge, Mass.: MIT Press, 1985.

McLuhan, Marshall. *Understanding Media*. 1964. Reprint, Cambridge, Mass.: MIT Press, 1994.

Patterson, Freeman. *Photography and the Art of Seeing*. Toronto, Canada: Key Porter Books, 1989.

Ricoeur, Paul. *Time and Narrative*, vol. 1. Translated by Kathleen McLaughlin and David Pellauer. Chicago: University of Chicago Press, 1984.

Rybczynski, Witold. *Home: A Short History of an Idea*. New York: Viking Penguin, 1986.

Scarry, Elaine. *Dreaming by the Book*. New York: Farrar, Straus and Giroux, 1999.

States, Bert O. *Dreaming and Storytelling*. Ithaca, N.Y.: Cornell University Press, 1993.

Stroupe, Craig. "Visualizing English: Recognizing the Hybrid Literacy of Visual and Verbal Authorship on the Web." *College English*, May 2000.

Todorov, Tzvetan. *Genres in Discourse*. Cambridge, England: Cambridge University Press, 1990.

Venturi, Robert, Denise Scott Brown, and Steven Izenour. *Learning from Las Vegas*, rev. ed. Cambridge, Mass.: MIT Press, 1977.

Zizek, Slavoj. *The Plague of Fantasies*. London: Verso, 1997.

—Robert Atwan

Contents

6 REDEFINING MEDIA 123

Teaching *Convergences:* An Introduction

ESTABLISHING CONTEXT

Throughout history, the emergence of new media has led to conflicting cultural viewpoints. In ancient Greece, the development of alphabetic literacy was deemed a boon by those who saw its potential for the broad dissemination of scientific, mathematical, philosophical, and literary knowledge. Yet not everyone agreed. Somewhat ironically, one of the Socratic dialogues Plato wrote attacked writing. Through the voice of Plato, Socrates expressed a distrust of writing as a means of exploring philosophical truths, which in his opinion could be done only through face-to-face conversation. Centuries later, some saw the development of photography as a miraculous advance over painting for capturing or documenting reality, while others recoiled in terror at each click of the camera's aperture, convinced that the demonic box had stolen a piece of their soul. Advocates of television thought the new visual medium would usher in a new democratic age wherein the common citizenry could take control of the airwaves, but dissenters criticized the boob tube's ability to hypnotize viewers and turn them into mindless couch potatoes. Similar disputes are occurring today around the potential of the Internet—is it truly the great democratic equalizer of our time, as some claim, or is it simply a tool for highly targeted corporate marketing (not to mention a major venue for pornographers)?

Though certain factions in our society (academics, politicians, spiritual leaders) continue to question their merits, we have for the most part gotten used to having new communications media and technologies in our world. Radios and television sets can be found in most American homes, massive bookstore chains anchor our shopping centers, and the percentage of people who have yet to "log on" is growing smaller every day. Your students are a part of this scene, yet they may not have questioned the presence of mass media in their lives because it is so ubiquitous and pervasive.

Convergences, then, can be used to guide their explorations into the complex relationships between these various media and the culture they constantly reshape. As instructor, you might think of yourself as a multimedia guide who poses the questions necessary for self-reflexive investigation: How does the medium of a text influence our interaction with it? What kinds of cultural associations do we carry into our readings of a text? Is there a "history" to how we read texts that has been influenced by the presence and dominance of particular media? How are we, as readers or viewers of texts, constructed by those very texts?

A NOTE ON USING THE EDITORIAL APPARATUS

Convergences contains examples of texts from many different media, grouped by thematic content rather than by form. In addition to the written texts common to most composition readers—poems, articles, and essays—there are also visual pieces such as paintings, cartoons, photographs, and Web pages that serve to expand our notion of what constitutes a text in our age of mass media. Then there are multimedia selections—photomontages, weblogs (or "blogs"), the "uncommercials" concocted by Adbusters, Nam June Paik's mixed-media video installations—that show how the boundaries between different media are easily blurred. The book's purpose is to allow students to explore the common means by which writers, photographers, designers, painters, and others use the languages of shape, color, layout, graphics, and words to create their original texts. In so doing, *Convergences* asks students to consider the relationship between a text's content and its form; in line with this goal, discussion and writing assignments alike should consider not just what each text says but also how, why, and for whom it is said.

Convergences uses a three-part structure: Accompanying each selection are questions that deal with the *message* conveyed in each text, the *method* by which that message is conveyed, and how the choice of *medium* affects these other categories. More than just catchy and alliterative, the Message/Method/Medium apparatus is designed to show how these features are intimately related to one another in virtually any public text, from the front page of a traditional newspaper to a novice's personal Web site. You should direct student discussion in ways that complicate the divisions of these three categories, leading them to see how (to quote Marshall McLuhan) the medium is the message. For example, if the very same text were placed on both a newspaper page and a personal Web site, how would our reaction to it be influenced by the two media? Would we be more trusting of the text embodied by the paper because of our cultural assumptions regarding the newspaper's relative authority? The Message/Method/Medium questions also show how different media are capable of transmitting similar messages; not only do individual selections contain texts from different media (a Joyce Carol Oates story and an Edward Hopper painting), but the book's six broad themes range from concepts of self-identity in "Staging Portraits" (Chapter 1) to notions of the rhetorical strategies employed when "Making History" (Chapter 4). Further underscoring these intertextual connections are the end-of-chapter questions, Writing/Researching/Collaborating, which you might consider incorporating throughout your exploration of the chapter.

As another component of the book's apparatus, the Key Terms entries offer instructors an invaluable way to incorporate writing instruction into a discussion of *Convergences*'s various themes. The highlighted terms are rhetorical concepts that are illustrated in the chapter's contents—"metaphor" or "point of view," for example. To make the most of this feature, you may want to adapt your writing exercises so that they coincide with the introduction of these terms and concepts—for instance, have students focus on the shark as a metaphor when they read Terry Tempest Williams's meditation on Damien Hirst's shark installation.

ADDITIONAL APPROACHES

Finally, consider the ways in which you and your students can arrive at a shared lexicon for discussing these selections; this will allow students to see how people working in different media can use similar rhetorical strategies, a connection that should ultimately be drawn to the writing that students themselves will produce. How is metaphor crafted in a poem, a painting, a commercial? What specific narrative techniques are used to construct a prose essay as opposed to a comic book? Can a newspaper article be "cropped" the same way a photograph can? Also, think of ways that you might incorporate the notion of translation into your writing assignments or discussions. By asking students to consider how a Gwendolyn Brooks poem could be rewritten as an essay or how a Gloria Steinem article could be retooled into a documentary film, you ask them to think about what is gained and/or lost by putting a text into a different medium. Such exercises will also enable students to develop a better understanding of the conventions and aesthetic features that differentiate these media. As you foster your students' writing and critical thinking abilities and help them gain confidence as writers of arguments, it is important to remind them that every selection in *Convergences* is the product of a person or people making choices designed to elicit specific rhetorical effects—from which medium to work in to which type of camera lens to use, from a choice between two words to a choice of how to structure an essay. These very same choices are at their disposal, too.

READING TEXTS

When approaching any type of text, regardless of its medium, you should begin analyzing it with certain basic questions in mind:

- What media are used in this text, and what generic conventions can be identified?
- Where is the "eye" of the text? In other words, what point of view did a short-story

writer choose, where did a photographer position the camera, or what perspective did a painter decide on?

- What is the focus of the text? What is put in or left out of the text's frame or scope? This includes material objects as well as such elements as lighting, shading, background, and soundtrack.
- How is this particular text edited? How are time and space manipulated in this text (flashbacks, jump cuts, voice-overs, etc.)?
- In what format did the text originally appear? Did an article appear in the *National Enquirer* or in the *New Yorker*? Was a movie shown in a small arthouse theater or a twenty-four-screen multiplex?
- For what audience is the piece intended? What position is assumed by the author of the text? What position does the creator assume for the reader/viewer?
- What role does intertextuality play in our understanding of the text? What other cultural texts or contexts do we need to be aware of in order to "get" the message behind a particular text?

The following groups of suggestions offer possible strategies for helping students approach the reading of texts from a variety of media and genres.

READING AN ARTICLE OR ESSAY

Like any other text, an essay should be approached with a dual purpose: identifying the argument or message as well as the structure of the piece. Students should be prepared to answer questions such as the following:

- What happens in the course of the essay? What stance is adopted at the beginning, and how does that stance change by the essay's end?
- What tone or voice does the author use? What do word choice and figurative language suggest about the essay's overall message?
- What relationship is assumed between writer and reader? Are readers already in league with the author, or must they be convinced to accept the author's position?
- Is the essay's purpose either to inform, to persuade, or to entertain? Or does it have multiple purposes?

READING A POEM

Reading poems can be a daunting, complicated task. Poetry has a complex system of formal rules, and poets often reach beyond the limits of everyday language. While you might not want to subject your students to studies in scansion and meter, they should be able to do the following:

- Identify the layout of the poem. How are lines broken on the page? Is it separated into different stanzas? Does this serve to emphasize certain words or images? A particular rhythm? Reading aloud can often illustrate these features.

- Identify the poem's speaker (who is not always the same as the poet). What clues in the poem tell us about the narrator?
- Point out striking imagery and figurative language. Sometimes even the smallest of details work together to create a singular impression of an idea, an event, or a person.
- Recount the plot (if there is in fact a plot). What happens in the poem? Are there specific characters or settings in the poem?

READING A SHORT STORY (OR PROSE FICTION EXCERPT)

The formal elements of short fiction generally include a distinct narrative that has a basic plot, characters, setting, and a point of view that students should be able to identify. Questions to ask include the following:

- What is the story's plot? What is the temporal "map" of the story, and what changes or transformations (if any) occur during the course of the narrative?
- Who are the characters in the story? Who acts, and who is acted upon? Are we meant to identify with some characters over others?
- How does setting affect the story's message? For instance, does a certain place emphasize the main character's moral stance, or does it work against him or her?

READING A WEB SITE

Though a fairly recent medium, the Internet makes use of many established verbal, aural, and visual media, incorporating their elements into Web sites with a variety of purposes. In addition to the basic formal questions listed above, you might also have students consider the following:

- What is the ratio of verbal text to images, sounds, and other media? What does this suggest about the site's purpose? Is a flashy site more entertaining, engaging, or informative than a static one? What do these things say about the site's intended audience?
- What level of interactivity is built into the site? Are we meant to simply sit and read, or are we encouraged to point and click, leave the boundaries of the site, download video clips, and so on? What does this suggest about the respective levels of control between site and navigator?
- What types of hyperlinks are included on the site? Are we urged to click through to Amazon.com and buy a book, or are we sent to a political activist site to find out more information on a topic?
- How is the information architecture structured? How big is the site, and how is the content divided up among pages? Is the structure hierarchical (suggesting that certain information is more important) or parallel (suggesting equal status)?

- What navigational cues are given to the user? Does the design offer clear, directive signposts, buttons, or menu options? Or does it advocate exploration by hiding links, using rollover graphics, and so on?

READING AN ADVERTISEMENT

In helping your students become more critical readers of advertisements, it is important to extend their notion of an ad's purpose beyond that of bottom-line profit. Pairing legitimate ads with the spoof advertisements included in this book (those by Barbara Kruger or Adbusters, for instance) will help deconstruct some of these strategies. The following are some additional questions you might consider:

- What is actually being sold in the ad? The product itself, or an ideology or image that the advertisers wish to associate with the product?
- Does the ad appeal to the audience's emotions, sense of logic, or the company's reputation? Or is a combination of these appeals used instead?
- What is the most important element of the ad? Text? Graphics? A catchy slogan or jingle? The company's logo? When reading, pay attention to where the eye is drawn initially, which pieces of the ad stand out.

READING A MONUMENT OR SCULPTURE

When reading a sculpture, monument, or installation, it is important to keep in mind that the piece's artistic meaning is often subject to the interplay of several different media (Nam June Paik's video installations interact with natural elements like plants, and Maya Lin considered the contours of the surrounding landscape when designing the Vietnam Veterans Memorial). Also keep in mind the following:

- What materials are used in the piece, and what associations are embodied in them? A choice between black granite and white marble may at first seem arbitrary but can indicate the kind of tone the artist wishes to convey.
- Is the subject portrayed in an abstract fashion or a realistic one? Does this choice reflect cultural attitudes about the subject matter or challenge them?
- Where is the artwork meant to be seen? A museum exhibit? A mall? A park? What does this choice suggest about the work's purpose and its level of seriousness?

READING A PAINTING OR PHOTOGRAPH

The generic conventions engendered in these two media are not as closely related as might be assumed. Historically, painting has been associated with high art, while the relatively new technology of photography has fought to gain that status. Photography's value as a medium is that it ostensibly has the power to show us events that actually happened. Still, there are similarities between the two media. Questions you might ask:

- Is the painting or photograph meant for a particular context? Is it a work of art, or does it serve a commercial purpose?
- How are color choice and composition used to convey emotion or symbolism? How does black-and-white photography differ from color photography? For instance, does black-and-white suggest truth, simplicity, or nostalgia?
- Is the subject matter abstract or representational? That is, does the subject matter represent something we know, or is it completely invented? What does this choice tell us about the idea behind the piece? With respect to photography, why might an abstract treatment be chosen; with respect to painting, why might a realistic perspective be chosen?
- What is the scale of the work? What might this suggest about the subject's degree of importance in the eyes of the creator?

Of course, these lists of questions are neither exhaustive nor exclusive to the media under which they are grouped. Be sure to adapt these questions for use in exploring other media as well.

DIVERGENCES

The following texts and Web sites contain additional information about media studies, visual literacy, and pedagogical approaches to these subjects.

Print

Barnard, Malcolm. *Art, Design, and Visual Culture.* New York: St. Martin's Press, 1998.

Berger, John. *Ways of Seeing.* London: Penguin Group, 1972.

Bolter, Jay David, and Richard Grusin. *Remediation: Understanding New Media.* Cambridge, Mass.: MIT Press, 2000.

Walker, John, and Sarah Chaplin. *Visual Culture: An Introduction.* New York: St. Martin's Press, 1997.

Welch, Kathleen. *Electric Rhetoric: Classical Rhetoric, Oralism, and a New Literacy.* Cambridge, Mass.: MIT Press, 1999.

Web

http://db.education-world.com/perl/browse?cat_id=1135. The *Education World* Web site features lesson plans, activities, and other resources for teaching visual arts and humanities.

http://vos.ucsb.edu/shuttle/media.html. "Voice of the Shuttle: Web Page for Humanities Resources" has links to several media studies sites.

http://www.com.washington.edu/rccs/. The Resource Center for Cyberculture Studies (RCCS) is a nonprofit support organization for teachers, students, and other explorers of emerging cybermedia.

1 Staging Portraits

The portraits contained in this first chapter are surprisingly diverse, ranging from the traditional Americana of Norman Rockwell to Ellen Harvey's paintings of (fake) Polaroid photographs of people making "monster" faces to Christine Castro's maganda.org, which, through image, word, and hypertext, serves as a multifaceted, ultramodern self-portrait. The question these portraits ask, and the question you should ask your students as they begin this chapter, is "What is a portrait?" Students will answer this in a variety of ways throughout the chapter.

The topic of this chapter suggests an excellent way to get students to write: as a beginning exercise, preview with students the different types of portraits the chapter contains. Keeping the various definitions of "portrait" in mind, have students produce their own "self-portraits," describing their physical characteristics, their likes and dislikes, their family, or anything else they find appropriate. By focusing on themselves as subjects, those who are otherwise reluctant to write may open up. Also, the numerous types of portraiture presented in the chapter may be encouraging for students who feel lost when asked to describe themselves in writing.

SELF-PORTRAITS
Cindy Sherman

2 PHOTOGRAPHS

Much of the popular perception of photography is surprisingly simple: what you see is what you get. The photographs here challenge that response. Cindy Sherman's approach to art involves commenting on the medium itself in order to remind the viewer of photography's artificial, crafted nature—what you see is quite often not what you get. As an example of postmodern art, Sherman's commentary also extends beyond the medium; students might benefit from a discussion about postmodernism's characteristic gestures of self-reflection, parody, and intertextuality.

Despite their apparently uncomplicated style, Sherman's pictures can be seen as complex critiques of familiar cultural representations. Her works challenge cultural constructions of femininity, sexuality, celebrity, and power by addressing them ironically—her "film stills" are, of course, not real film stills, and the subjects she portrays are not movie actors. Ultimately, Sherman's pictures challenge the act of looking itself, as she is simultaneously the subject of her photos as well as the figurative eye behind the camera. Feminist film theorist Laura Mulvey, in a 1991 article for the *New Left Review*, offers this quote, which encapsulates Sherman's artistic vision:

> She is artist and model, voyeur and looked at, active and passive, subject and object; the photographs set up a comparable variety of subject positions and responses for the viewer. There is no stable subject position in her work, so the Film Stills' initial sense of homogeneity and credibility break up into the kind of heterogeneity of subject position that feminist aesthetics espoused in advance of postmodernism proper.

To help contextualize Sherman's work, you might present students with examples of the "straight" texts to which the photos refer. For instance, bring in pictures of Elizabeth Taylor, Bette Davis, or a similar 1950s film ingenue to pair with *Untitled Film Still #48*, and a more provocatively posed model to stand beside *Untitled #96*. You may also want to point out the thematic variety in Sherman's work: the seemingly playful commentary on cinematic themes in *#48* and the more disturbing answer to pornographic portrayals of women in *#96*. You could emphasize the artist's progression toward graphic, disturbing imagery by examining some of her later efforts from the 1980s and 1990s.

MESSAGE

In each photograph, Sherman portrays herself as a unique persona or character. The first selection may remind some students of film noir—Hitchcock films, for example. Asked to compose narratives around the photographs, students might play with genre-specific conventions (a woman on the run after being framed for a murder she didn't commit, finally to be vindicated by a hard-nosed, down-on-his-luck private eye). The second selection should remind students of spreads featured in magazines such as *Maxim* or *Playboy*. Given the photograph's take on the centerfold, and students' likely familiarity with this style, the piece should spark lively discussion.

METHOD

Although the pose, costume, and lighting of *#96* are similar to those of the centerfold genre, Sherman's gaze—aloof, disconnected—starkly rejects the assumed erotic nature of such a photograph. The dimensions of the photo—nearly human-size—amplify this point by discomfiting the viewer.

MEDIUM

Sherman's photographs allude to larger works that do not properly exist. Despite that, however, they are constructed in such a way that the artist/model is the protagonist, even if she is the only figure in the shot. She is "acting" in a self-aware way, and the audience's participation creates the notion of "performance."

ADDITIONAL WRITING TOPICS

1. Reflect on a time when you sat for a portrait (with family or at school, for instance). Describe the scenario by focusing on the elements that uncover the staged nature of the portrait. How were you posed? What props or backdrops were used? What sort of conversation occurred between you and the photographer?
2. Your instructor will supply the class with several random photographs. Choose one and construct a narrative about what items might be out of the picture's frame and why the photographer chose to crop the shot as he or she did.
3. Many critics consider the shape-shifting persona of pop icon Madonna to embody many of the themes found in Sherman's art, particularly those dealing with female identity. Find examples of Madonna's various images online or in print publications, and compare them to Sherman's film stills, indicating similarities and differences.

DIVERGENCES

Print

Goodrich, Norma L. *Heroines: Demigoddess, Prima Donna, Movie Star*. New York: HarperCollins, 1993. This collection of essays offers an explanation of the cultural relevance behind the female icons that Sherman's art attempts to deconstruct.

Mulvey, Laura. "Visual Pleasure and Narrative Cinema." Screen 16, no. 3 (1975): 6–18. Reprinted in *Visual and Other Pleasures*. Bloomington: Indiana University Press, 1989. This important feminist film theory article is one of the first to argue that the symbolic logic involved in viewing cinema constructs what Mulvey calls the "male gaze," meaning that the viewer being assumed by the camera's vantage point derives distinctly masculine pleasures from looking.

Web

http://www.eng.fju.edu.tw/Literary_Criticism/feminism/index.html. Fu Jen, Catholic University's Feminisms and Gender Studies site, offers several images dealing with pictorial representations of women, including works by Cindy Sherman and Barbara Kruger, as well as photos of Madonna and the Spice Girls.

Audio/Visual

Office Killer. Directed by Cindy Sherman. 83 minutes, rated R. Dimension, 1997. Videocassette. Sherman's recent attempt at actual filmmaking recounts a female copyeditor's descent into madness and makes for an interesting counterpart to Sherman's film stills; here, students can see Sherman's artistic attitude toward the medium of cinema fully realized.

Portrait of an Artist at Work: Cindy Sherman. Directed by Michel Auder. 1988. Videocassette. This documentary shows Sherman at work and explains some of the meaning behind her art.

What Did You Expect?
Dorothy Allison

ESSAY

Like the Cindy Sherman selection, Dorothy Allison's "What Did You Expect?" lets its readers behind the scenes in order to reveal the world of photography's nonrealistic nature. Allison suggests that the photograph does not allow people to see her true self; the proper medium for that task is her writing. It is through writing that Allison can explain why she chooses to present herself as she does, why she resists depictions that she finds insulting. The static photograph is not capable of moving through time to show us the influence of Allison's heavily made-up mother and sisters or the shame she felt being associated with redneck culture.

Allison's problem is not exactly one of representing her true self, though. She admits that she has carefully constructed her own individual persona over the years, a composite of positive and negative events from her past. She writes, "My persona is as much a conscious rejection of my mother's armored features as it is an attempt not to cater to the prejudices and assumptions of a culture that seems not to want to look at women like me. It is not seamless, merely stubborn."

Beginning the discussion of this text with a consideration of the concept of "persona" may prove challenging but valuable, especially if students describe the ways in which they construct their own personae and the reasons behind those choices. After this, explore comparisons to Sherman's work. Consider the ways in which both artists use and comment on composition—this could lead to a discussion of such themes as irony and parody. Allison's point, in part, is to resist parody so as not to become a victim to caricature. However, consider that her eventual decision to pose in a Laundromat instead of being covered in powdered sugar could be seen less as a victory and more as a compromise. Finally, analyzing Allison's consistent choice of figurative language promises to be a worthwhile exercise in understanding her arguments. For example, in her use of the word "armor" as a metaphor for makeup, Allison (like Sherman) indicates the oftentimes constructed and protective nature of feminine personae.

MESSAGE

"Powdered sugar" is a reference to the South's cultural heritage, which (especially among lower and working classes) involves rich foods laden with sugar, butter, lard, or bacon grease. For this reason, Dorothy Allison sees the photographer's proposal to cover her in powdered sugar as a cliché. The final picture is in some ways a lesser version of the sugar concept, but she resigns herself to it—though not without a twinge of guilt brought on by being among other working-class women actually doing their laundry.

METHOD

Students will be quick to note that Allison isn't altogether thrilled with the prospect of being photographed—a fact she readily admits. Partially, she wants to resist the same types of social pressures that caused women like her mother and sisters to don "war paint" so they could take on the world. Also, visual representations of the working/lower class are often embarrassing depictions of shame-filled arrests, scandals, and tacky affairs. Allison desperately wants not to participate in these types of representations. Holding some position of celebrity, she doesn't want to betray her working-class background by making a joke of it. The piece's title might be taken in a couple of ways, depending on whom it is meant to address: if the "You" in the title refers to the photographer, it could represent a challenge to the photographer's assumption that Allison would comply with her somewhat insulting request; if the "You" is the reader of the piece, it could represent a knowing affirmation of our lack of surprise at the photographer's clichéd suggestion.

MEDIUM

Just as Allison prefers to dress for comfort and feels no need to freshen her makeup as she deplanes, her writing style suggests a similar attitude—no-frills, honest, entirely invested in maintaining a sense of authenticity. Allison seeks to challenge even her own physical actions by offering a confessional voice that reflects on those actions. For instance, she is able to ascribe an antagonistic meaning to her smile, which she inherited from her mother; she is able to describe her slight discomfort at being talked into posing in a Laundromat; and she even confesses to her own naive idealism as a young woman who aspired to be a banner-carrying hero of feminism.

ADDITIONAL WRITING TOPICS

1. Visit a local bookstore and look for hardcover versions of books by several popular authors (Stephen King, Tom Clancy, Anne Rice, or Douglas Coupland, for example). Write a comparative essay analyzing the dustjacket photographs of these authors. Do the poses and props suggest things about the writers? Are the images in keeping with any preconceptions you may have about the personalities of these writers?
2. Imagine you are a professional magazine photographer assigned to shoot Dorothy Allison for an upcoming interview. Assuming the editor gave you a copy of "What Did You Expect?" prior to the assignment as background material, describe your own approach to the photo shoot—how would you act differently from the photographer in the essay? Give reasons for your decisions.
3. Allison describes her reaction to Fox Entertainment's *Cops* as a mixture of shame and outrage, noting: "The first time I saw that television show, I sat through the whole thing with my mouth hanging open, unable to look and barely able to stand what I was seeing." Watch a show such as *Cops* or *The Jerry Springer Show*, and then compile a list of characteristics that help make up the "white trash" stereotype. Make up an alternate list of the positive traits offered in Allison's article.

DIVERGENCES

Print

Allison, Dorothy. *Skin: Talking about Sex, Class, and Literature*. New York: Firebrand, 1994. This collection of short essays and articles covers the topics suggested in its title.

Web

http://www.salon.com/books/int/1998/03/cov_si_31intb.html. In this *Salon* interview, Dorothy Allison discusses coming to terms with her identity as a lesbian, a mother, and an acclaimed writer with a working-class upbringing.

Audio/Visual

Bastard Out of Carolina. Directed by Angelica Huston. Starring Jennifer Jason Leigh, Christina Ricci, and Lyle Lovett. 101 minutes, rated R. Fox Lorber, 1996. Videocassette. This film adaptation of Allison's much-celebrated novel of the same name shows her heroic treatment of the working-class South.

"Dorothy Allison." Distributed by WHYY, Philadelphia, March 23, 1998. Recording. This National Public Radio broadcast of *Fresh Air*, hosted by Terry Gross, is an interview with Allison about *Bastard Out of Carolina* and her other publications.

maganda.org
Christine Castro

3 SCREEN SHOTS

Christine Castro exemplifies an entire subculture of people daring to publish online the type of writing usually hidden from the eyes of others in personal, locked diaries. Her weblog, or blog, offers students an appropriate illustration of the collapse of public and private space in our contemporary society, and they may want to consider what it is about the Web that entices certain personalities to air their dirty laundry to complete strangers. Does this new medium simply make it easier to publish texts, or is some other cultural force at work here?

Navigating maganda.org is not an immediately intuitive experience. The links are purposefully vague, leaving the user with no idea of what exactly to expect. Students might benefit from accessing this site as a group so that they can share different ways of exploring it.

The content is very personal, which allows the user to feel familiar with Castro. Her voice is casual and nonthreatening, as is the site design, in spite of its exploratory architecture. This mood is consistently sustained throughout the entries and is carried out further in the color themes and layouts, which make good use of pastels, close cropping, fragments of text, and abundant white space. We get a sense of Castro's talent as a writer and visual artist, but these points are understated on her site; nothing is too flashy. In one sense, the site is extremely unified; the overall design, written content, and photographs have an unpretentious quality that students could spend some time analyzing. When viewing the two screen shots, the class should be able to point to various design elements and metaphors that maintain the site's continuity; students should also notice that the third screen shot stands noticeably apart from the others.

Finally, the blog phenomenon itself would serve as a good topic for discussing the variety of content on the Internet, as well as the variety of quality in that content. Is the personal blog just a narcissistic project for a new, bored generation? Or does it instead offer individuals a forum in which they can make their identities public and, in so doing, resist the homogeneous tendencies of commercialism?

MESSAGE

Christine Castro's site design gives us a sense of a Web-savvy woman who understands the associational logic of the Internet. Her use of images is in tune with the Internet's predominantly iconic, graphical landscape; accordingly, she writes in a quick, fragmented, Web-friendly fashion. The content is much like that of a traditional journal, but the Internet allows for other artistic media besides writing; additionally, Castro is able to provide hyperlinks to places beyond her site, giving the visitor a glimpse into a blogger community that extends far beyond one woman.

METHOD

Students might notice that Castro's pictures are very candid—mostly spur-of-the-moment, out-of-focus efforts. This aesthetic suggests a certain spontaneity, a particular sensitivity that is moved by beautiful images and experiences and attempts to capture them for others. This characteristic spontaneity is particularly well suited to the fast-paced, ever-changing feel of the Web itself.

MEDIUM

Maganda.org has a strong sense of identity, most evident in the use of color as a navigating element. The soft pastel blocks that designate the site's architecture are a good match with Castro's "sappy" content. Her use of paintings and drawings on the new site are meant to further convey a sense of the artist's identity, because these media, too, are creative outlets for her.

ADDITIONAL WRITING TOPICS

1. Analyze Castro's site in terms of its medium. What advantages to putting this content on the Web could not be gained by, say, keeping a handwritten journal? Are there certain drawbacks to being "published" on the Internet rather than in print? What effect is achieved when artwork in other media (paintings, drawings, photographs) become digitized on Castro's site?
2. After visiting maganda.org, construct a brief narrative of what a typical day in Christine Castro's life might be like. Consider events that she would make note of on her site as well as those she might choose not to disclose, and offer rationales for such choices.
3. Sign up for a Web account at Blogger.com and keep your own blog for a couple of weeks. Share your blog with the rest of the class, and then discuss why you chose to include the content you did.

DIVERGENCES

Print

Lopez, Erika. *Flaming Iguanas: An Illustrated All-Girl Road Novel Thing*. New York: Scribners, 1998. This book is among the many influences Christine Castro cites on maganda.org. The novel's playful typography, images, and narrative structure, as well as a story focused on the heroic deeds of strong women characters, are easily recognizable themes on Castro's site.

Web

http://dairyland.com. Compared to Blogger.com, this site offers a slightly more basic, beginner-friendly version of the blog portal concept.

http://www.blogger.com. This site offers subscribers free Web space and a template for creating their own blogs, as well as links to a truly varied community of bloggers.

http://www.webbys.com. The site for the Webby Awards shows a range of different Web-page designs. Students could discuss the possible criteria of good designs.

Audio/Visual

Hackers. Directed by Iain Softley. Starring Angelina Jolie, Johnny Lee Miller. 104 minutes, rated PG-13. MGM/UA Studios, 1995. Videocassette. This marginal thriller about the hacker subculture and adolescence could be used to supplement discussions about whether or not there remains a gender divide in digital culture. One of the film's protagonists—played by Angelina Jolie—is especially adept as a hacker.

MONSTERS
Ellen Harvey

9 PAINTINGS

Ellen Harvey is a serious painter whose artistic project is, ironically, to save painting from taking itself too seriously. Harvey has mentioned that she aims to show how painting needs to evolve beyond the depiction of the fantastical and the impossible; that's now a job for cinema. Instead, she wants to show how painting can be personal, local, and accessible. "It's cheap. Anyone can do it," she says. She wants her audience to see painting in new ways; as the title of her current, ongoing series indicates, she wants us to conceive of paintings as *Low Tech Special Effects*. She does this by referencing the banal and ordinary Polaroid photograph. In spite of Harvey's very technical photo-realistic style, the Polaroid comes across as an odd subject matter for a "serious" art such as painting.

To gain a further point of entry into Harvey's paintings, you might want to start a discussion on the differences between high art and low art, as well as the media associated with each (refer to the Greenberg article cited in the Divergences section on p. 9). Students could consider how our culture ranks different forms of art such as painting, film, television, and comic books; they should also consider the relations between these rankings and socioeconomic divisions of high and low. A follow-up conversation could center on the ways in which Harvey's approach to painting, and art in general, blurs those divisions.

MESSAGE

The people posing for *Monsters* have broken the "fourth-wall convention" by acknowledging the presence of the artist—they make eye contact with the viewer and consciously contort their features. Given this playful tone, the paintings fail to be traditional portraits. They can, however, be read as self-portraits in the sense that the models chose their facial expressions. For instance, comparing Matty's expression to Alia's, one might get the impression that Matty has a more reserved personality, while Alia is less inhibited. This assumes, of course, that the paintings are based on actual pictures of actual people, a point about which Harvey is ambivalent. This ambivalence leads to the possible message of the series—what is real about portraiture, painting, or art in general, and what is artificial?

METHOD

The *Monsters* paintings' wrinkled white borders and careless, close cropping perfectly capture the spontaneous, disposable feel of the Polaroid aesthetic. (Students may be more current with digital cameras or with Polaroid's new I-zone camera, which uses a fast-developing sticky film.) The poses, too, are in keeping with the Polaroid look; discuss what types of occasions are appropriate for using this medium. Finally, point out that the scale and multitude of images in the original exhibit would have enhanced Harvey's intended effect, and individuals would matter less than the act of posing itself. Harvey's work, in part, attempts to show the humor involved in having her subjects pose as "monsters" with no props or technological assistance. Rather than resembling monsters, the results remind us of the well-known juvenile act of mugging for the camera, a crime of which many of us have been guilty.

MEDIUM

Harvey's paintings strive for photo-realism, which is ironic considering that they depict people in the act of trying hard not to be real. Unlike traditional attempts at photo-realism by American painters of the mid-twentieth century, Harvey's paintings are in part kitsch—they are high-art representations (paintings) of a low-art medium (silly Polaroid pictures).

ADDITIONAL WRITING TOPICS

1. One of Ellen Harvey's artistic works, entitled *ID Card Project*, is a collage of her various identity cards. Examine the pictures on the forms of identification you carry with you (your school ID, driver's license, or library card, for example), and construct a narrative about the days during which some of these pictures were taken.
2. In the headnote, Harvey refers to the Polaroid picture as a "nostalgic technology of representation." In what ways is the Polaroid nostalgic? What other technologies do you find nostalgic? Write an essay in which you recount your interaction with these technologies and explain why they have remained special to you.
3. See if you can locate old photographs of your friends and/or relatives. Describe some of these photographs, particularly the people in them, in as much detail as possible. How well do these pictures capture the identities of your friends and relatives as you know them?

DIVERGENCES

Print

Greenberg, Clement. "Avant-Garde and Kitsch." *Partisan Review* 1, no. 5 (Fall, 1939): 34–49. Greenberg's classic article outlines the modernist distinctions between highbrow and lowbrow art, distinctions that postmodern art such as Harvey's tries to do away with.

Web

http://www.ps1.org/cut/studioimg/harvey.html. The International Studio Program's site includes a detailed résumé of Ellen Harvey. Also included is an artist's statement and explanations of several of Harvey's recent exhibitions.

http://owa.chef-ingredients.com/postUK/20/harvey.htm. On this page, Harvey describes painting as "a low tech special effect" and expands that philosophy in a short paragraph accompanied by photos of Polaroid paintings from *Monsters* and *Into the Abyss*.

HOME MOVIES
Judith Ortiz Cofer

MEMOIR, BOOK COVER, POEM

One of the most interesting aspects of Judith Ortiz Cofer's writing is the close relationship it seeks to establish with the other media it references. With its shifts in voice and its abrupt jumps from past to present, the memoir "Silent Dancing" mirrors the short and choppy home movie on which it is based; by contrast, the poem "Lessons of the Past" achieves its effect by alluding to another photograph (of ostracized Parisian women following World War II) in order to establish a powerful if not disturbing metaphor. These relationships serve one of Cofer's consistent goals when she writes: to keep the memory of her childhood and her Puerto Rican culture alive through imagery that strives toward realism rather than sentimentality.

Several of Cofer's techniques would make for interesting writing assignments or discussion starters. For example, the narrative shifts (indicated by italics) in "Silent Dancing" take the reader from the "primary" past tense of childhood to the "secondary" present tense of the home movie. Students could not only discuss the effect of these shifts but also examine other techniques (e.g., tone and diction) that differentiate the memoir's parts.

The memoir and the poem alike call into question the reliability of memory; in "Silent Dancing," Cofer confesses to not being able to remember her childhood in color but in shades of gray, while the poem is based on an incident (falling into a fire) that is disputed by Cofer's mother. This recognition of the imperfection of memory could lead into an examination of how we judge the reliability of a particular medium for recording documents.

Cofer's translation of a photograph into poetry could also invite a discussion about the way we interact with photographs—we often see them as more than static images, perhaps even as links (similar to those in hypertext) to a vivid realm of remembrance. This discussion could refer back to the selections of Cindy Sherman and Dorothy Allison, where the photographer and the writer critique the idea that photographs can ever actually capture reality.

One productive direction in which to take class discussion is toward the topic of Cofer's expert use of description. Her metaphors, similes, and other tropes may serve as exceptional models—though the language is figurative, the intent is to somehow render the memory more real. The tastes and smells, the peripheral memories of 1950s television shows and grocery trademarks, and the use of the Puerto Rican phrases of Cofer's community all work together to create a texture of random associations similar to the process of memory.

A final topic for consideration involves viewing the memoir, the poem, and the photograph as different ways of capturing or recording memory. After examining the advantages and disadvantages of the various media (and genres), students could debate how successful each text's attempt is at capturing the remembered event, what aspects of memory are emphasized or de-emphasized in each, and which ones are better at conveying the memory to a viewer/reader.

CALLOUT QUESTION

It might be advisable to ask students to look at and describe the photo from the cover of her memoir before actually reading the poem. Though they might point to her posture, the forlorn expression in her eyes, and the doll-like costume as indicators that she was a momentarily unhappy child unwillingly posing for a photograph, a more encompassing, deep-seated sadness is strongly suggested in "Lessons of the Past." Specifically, Cofer mentions her cropped hair and large eyes as reminiscent of a photograph of Parisian collaborators, steeped in fear and shame, and put on display for others to see. The earnestness in her choice of metaphor is meant to explain the grave coloring of her childhood memory.

MESSAGE

Silence is symbolic of the shortcomings of memory. Although Cofer has a home movie to help her remember the past, it is incomplete, for it lacks sound. The silence also represents family secrets that appear in her dreams, such as her cousin's abortion. Cofer refers to the music that was often playing and the sounds of Spanish being spoken in neighboring apartments. For her mother, the sound of Spanish was a rare comfort in an unfamiliar country. Sounds for Cofer, then, represent a sense of belonging, of home, that images only partly convey.

METHOD

In "Silent Dancing," Cofer weaves the scenes from the home movie in between the straight narrative of her life in Paterson, New Jersey. This technique is effective in providing a vivid backdrop against which to fit the facts of the essay. She evokes the physical realities of that time through her descriptions of the clothing her family wore, their postures, and the color of the room.

MEDIUM

The film does not tell a story so much as it provides imagery for Cofer's storytelling. Cofer uses the voices of her cousin and aunt along with an anonymous voice, presumably that of her mother or grandmother. The voices fill in the gap left by the silent movie—they provide facts and descriptions of people and events alluded to in the film. Although they appeared in Cofer's dreams, and may or may not be "true," the narratives help us make sense of the film.

ADDITIONAL WRITING TOPICS

1. Did you or your family record important moments during your childhood, such as vacations, graduations, or Little League games? Was there a designated method of recording these events, such as using a still or video camera, a journal, or a newspaper scrapbook? Revisit some of these documents, either in person or through memory, and characterize your response.
2. Cofer's memoir is extremely detailed, drawing upon the entire sensorium in order to recreate the moment: smell and taste figure into her writing as importantly as sight and sound do. Construct a similar narrative of one of your earlier memories involving taste and smell, and concentrate on including details that involve these senses.
3. Rewrite Cofer's poem "Lessons of the Past" as a short story. Follow up this activity with a discussion about the various challenges involved in translating poetical language.

DIVERGENCES

Print

Cofer, Judith Ortiz. *The Latin Deli*. New York: W. W. Norton, 1993. This is one of Cofer's best collections in that it represents the entire spectrum of her writing talents: poetry, memoir writing, creative nonfiction, short fiction. The selections are thematically unified in the project of "telling the lives of Barrio women."

Web

http://dive.woodstock.edu/~dcox/ohenry/cofer.html. Montgomery College's page dedicated to Judith Ortiz Cofer is one in a series of authors' pages.

http://parallel.park.uga.edu/~jcofer/. Judith Ortiz Cofer's professional page as a faculty member at the University of Georgia will allow students to explore how Cofer presents herself professionally and compare that to the persona she projects in her creative writing.

CONFESSIONS
Anne Sexton

RÉSUMÉ, PHOTOGRAPH, POEM

People often read Anne Sexton's writing as a guide for making sense of her tortured life. Consequently, students could find a useful avenue for discussion in Sexton's biographical background, which "Résumé 1965" touches on. Seen in the context of her former modeling career, her severe bouts of depression, and her eventual suicide, Sexton's poetry appears to have served a therapeutic function far more important to her than mere artistry or celebrity. In many ways, writing poetry kept Sexton alive. You may want to discuss with your students the different purposes for writing—or making any art, for that matter—and ask if they have experienced any of these during their lives. A short writing exercise in which they offer their personal definition of art and recount periods when they resorted to artistic outlets could help in furthering discussion.

Since the pieces in this selection are rather short, you may consider having students first regard the photo and describe Sexton's personality based on it alone. Propose the idea that this image of Sexton is not, as Arthur Furst suggests, that of an

"Egyptian queen preparing for the afterlife," but rather that of a trained fashion model who knew how to be photographed and who retains her sense of mystique. Students may then be surprised when they read her résumé and poem, both of which offer conflicting images of Sexton, images incompatible with the self-confident persona of the photograph. In "Self in 1958," Sexton gives us an ironic image of a "self" that has no ability to act on its own—it is just a shell, a doll, subject to the whims of whoever chooses to pick it up. Sexton's "Résumé," by contrast, offers the reader a more resistant persona, one who admits her insecurities and weaknesses while also pointing out her awards, successes, and strength of character. An analysis of the different tones, voices, and imagery in each piece could lead into a follow-up discussion about the different ways in which a persona can be constructed—in Sexton's case, students could debate whether the photograph, the résumé, or the poem best represents the "real" Anne Sexton.

CALLOUT QUESTION

In the common sense, a doll is a synthetic or artificial version of a person, but Sexton's use of the word is somewhat paradoxical. She refers to herself as a flesh-and-blood version of a doll, something that should be made of plastic or porcelain. The fact that she is a real person makes her an artificial version of a doll; like a doll, she is incapable of properly feeling or remembering, and she is unable to act on her own. Others play with her and manipulate her actions. The imagery in the poem is supported by words and phrases such as "plaster" (line 2); "Shellacked and grinning person" (line 4); "nylon legs, luminous arms" (line 9); "counterfeit table" (line 13); and "cardboard floor" (line 18).

MESSAGE

Though "Résumé 1965" includes basic biographical data, educational background, and information about Sexton's publications, the fact that it is written in prose form is the first surprise. Additionally, she provides very detailed information about her lineage, more personal background than is appropriate for a résumé, and disclosures on all sorts of personal matters, from her inability to get along with her peers as a child to her nervous breakdown as a new mother. Though Sexton's intended audience has never been definitively established, it is curious that "Résumé" was never published in Sexton's lifetime. Did editors reject it? Was it never submitted? Did Sexton simply write it for her own amusement?

METHOD

Fairy tales were Sexton's primary means of escape from the disillusionment of reality. The unbridled imagination of the fairy tale was far more preferable to her than a real world filled with real people, many of whom Sexton found unapproachable, confusing, or simply not worth the trouble. As Sexton matured, fairy tales were replaced by poetry, for it too represents a world where imagination reigns. As the last line of "Résumé 1965" indicates, poetry is everything for her; through it, she is capable of revealing her true self. Sexton remarks that she can't give her poems, which are an extension of herself, "someone's face-lifting-job."

MEDIUM

Arthur Furst's photograph shows a smirking Sexton; we see in her image here the kind of cynical wit that crops up in her résumé with references to Edward III's mistress and to her disappointment with her education. The psychologically wounded Sexton, the one who admits to having had a nervous breakdown and compares herself to an inanimate doll, is entirely hidden in the photo and for the most part in the résumé, but this persona is much more up-front in "Self in 1958." Her writing seems to point out her perceived frailties, her anguish at being trapped in domesticity and objectified as a doll or a fairy-tale heroine rather than as a real person.

ADDITIONAL WRITING TOPICS

1. In the poem "Self in 1958" and in "Résumé 1965," Sexton offers a couple of striking metaphors for herself: a doll and a tin can. Why do you think she chooses these metaphors? What are the similarities or differences between them? Also, look at the picture of Sexton. Does this image fit your conception of the person responsible for writing these pieces?
2. Analyze the structure of "Résumé 1965": How does Sexton order her information? What information does she choose to provide? What shifts in tone and voice can you find? Using Sexton's structure as a model, write your own personal résumé.
3. Sexton's résumé is a personalized approach to a characteristically impersonal genre. Conversely, Sexton's confessional poem laments her lack of person. Write an essay defending one of these pieces as the "truer" depiction of Anne Sexton. Defend your choice with evidence from each text.

DIVERGENCES

Print

Sexton, Anne. *The Awful Rowing toward God*. Boston: Houghton Mifflin, 1975; London: Chatto and Windus, 1977.

———. *The Book of Folly*. Boston: Houghton Mifflin, 1972; London: Chatto and Windus, 1974.

———. *The Complete Poems*. Boston: Houghton Mifflin, 1981.

———. *The Death Notebooks*. Boston: Houghton Mifflin, 1974; London: Chatto and Windus, 1975.

———. *45 Mercy Street*. Edited by Linda Gray Sexton. Boston: Houghton Mifflin, 1976; London: Martin Secker and Warburg, 1977.

———. *No Evil Star: Selected Essays, Interviews, and Prose*. Edited by Stephen E. Colburn. Ann Arbor: University of Michigan Press, 1985.

———. *Words for Dr. Y.: Uncollected Poems with Three Stories*. Edited by Linda Gray Sexton. Boston: Houghton Mifflin, 1978.

Web

http://home.sprynet.com/~mersault/sexton/. The Anne Sexton Bibliography is a regularly updated and very thorough compilation of primary and secondary sources, with several links to other sites.

http://www.coolmemes.com/reader/sexton.htm. The Cool Memes site has a large number of links to online texts, interviews, and reviews by or about Sexton.

http://www.levity.com/corduroy/sexton.htm. This fan site has links to audio files of some of Sexton's readings.

Audio/Visual

Girl, Interrupted. Directed by James Mangold. Starring Winona Ryder, Whoopi Goldberg, Angelina Jolie. 127 minutes, rated R. Columbia/Tristar Studios, 2000. Videocassette. Based on Susanna Kaysen's acclaimed memoir of the same name, this feature film deals with many of the themes in Sexton's writing.

Voices and Visions: Sylvia Plath. 60 minutes. Winstar Home Entertainment, 1999. Videocassette. A documentary of Sexton's close contemporary, whose life played out remarkably similarly to Sexton's own.

Betty
Gerhard Richter

PAINTING

In one of Gerhard Richter's journal entries from 1990, the artist ponders the fate of art since the modernist "readymade," the best-known example of which is Marcel Duchamp's 1917 piece *Fountain*, which consisted of an inverted urinal. Richter wrote: "Since then painting no longer represents reality but is itself reality (produced by itself). And sometime or other it will again be a question of denying the value of this reality in order to produce pictures of a better world (as before)." It may be helpful to provide students with background on the modernist art movement of the early twentieth century, which tried to free painting from having to refer to things in the real world and founded the philosophy of "art for art's sake" that gave rise to abstract expressionism and other nonrepresentational art.

Betty, like much of Richter's representational art, struggles with the very shaky line between an art that is returning to depict reality once again and one that strives to maintain its absolute freedom. For one, the painting is at three removes from reality—it is a painting of a photograph that has undergone certain artistic embellishments (age enhancement). Also, Richter's photo-realism has often been described as having abstract qualities that betray the realistic effect a viewer might expect—hazy lighting effects, unnaturally blending colors, abstract backgrounds. The curators of the Saint Louis Art Museum, by placing this painting beside one of Richter's gray paintings, apparently tried to emphasize the relationship between these two styles. It might, then, be a useful exercise for students to view several of Richter's paintings from different periods in his career. The move away from pure abstract art toward representation shows Richter trying to solve a philosophical problem through his art, a problem that is taking him several decades to work through.

MESSAGE

One of Richter's goals is to indicate how painting is different from photography, how it takes on certain duties that photography cannot. As noted above, the painting is an "age-enhanced" reproduction of a photograph of Richter's daughter, and it alludes to another of the artist's abstract works in the background. Both of these details stress the very materiality of painting by referring to itself and its medium (the common art-critic term is "self-referentiality"). Further, we might read the girl in the painting as looking at the painting behind her; even though her upper torso is turned, her lower torso remains fixed forward. In a metaphorical sense, the painting could be studying its own existence.

METHOD

The viewer is not given a direct portrait of Betty, but rather a portrait of the *idea* of Betty (which might make for a more captivating artistic statement because the viewer doesn't see Betty's face and must conjure one). This gesture, as some critics of Richter's work have suggested, is meant to make us focus more on the medium itself than on the things depicted inside it.

MEDIUM

Richter's painting is unusual in its ironic portrayal of the subject. In painting his daughter's turned head rather than her face, he renders the personal impersonal. The painting is made to look like a photograph through its use of distinct lines, a proper photographic perspective, and attention to realism.

ADDITIONAL WRITING TOPICS

1. Based on what you actually do see of Richter's daughter in the painting *Betty*, write a short description of her. How do you think she might look? What type of personality do you think she has? Does her pose in the painting suggest to you anything about her relationship with her father?
2. The common conception of a portrait is that it usually focuses on the face of the subject; if we think of *Betty* as a portrait, this isn't the case. How would you want your portrait done if you were not allowed to include your face in it? Consider such decisions as the medium, your pose, the lighting, the background, props, and any other elements. Explain how these choices help convey your personality.
3. Write an aesthetic response to Richter's painting. Does this type of artwork appeal to you? Defend your response by pointing to various features of the painting—the color palette, the overall composition, the choice in subject matter—that lead to your reaction. If you like the painting, what other favorite pieces of art (if any) does it remind you of? If you do not like it, what would make it better?

DIVERGENCES

Print

Buchloh, Benjamin. "Interview with Gerhard Richter." In *Gerhard Richter: Paintings*, edited by I. Michael Danoff, Roald Nasgaard, and Terry A. Neff. London: Thames and Hudson, 1998. This interview touches on subjects such as iconography in photography, Richter's thoughts on abstract painting, and the rhetoric of his art.

Richter, Gerhard. *The Daily Practice of Painting: Writings 1962–1993*. Edited by Hans-Ulrich Obrist. Cambridge, Mass.: MIT Press, 1995. This volume contains private journal entries and interviews with the artist.

Web

http://www.artchive.com/artchive/R/richter.html. The Artchive site has a few reproductions of Richter's purely abstract work from the late 1980s, with good resolution.

http://www.diacenter.org/exhibs/richter/richter.html. In the mid-1990s, New York's Dia Center for the Arts housed an exhibition for Richter called *Atlas*, a collection of over four thousand photographs, illustrations, and other small mixed-media pieces collected by the artist and used as subject matter for his paintings. Several images are included on this site.

http://www.dieter-obrecht.com/richter/richter.htm. Wildbrush's art.to.day is an "unofficial homage site" by a fan of Richter's work, with many of his paintings from across his career.

Audio/Visual

Sonic Youth. *Daydream Nation*. Blast First/Enigma Records, 1988. Album. The New York–based noise rock band Sonic Youth incorporates two Richter paintings (both titled *Kerze*, from 1982 and 1983) in its album cover design. It might prove an interesting exercise to have students listen to selections from the album or research Sonic Youth online, and then speculate on what image the band tried to convey by using Richter's paintings.

Bobland
Robert Atwan

MAP

Robert Atwan asks us in the headnote if we are able to imagine a map as a self-portrait, and in *Bobland* he provides us with an illustration. This whimsical exercise in wish fulfillment is Atwan's attempt to show us his vision of the "perfect place." As students discuss the map's function as a symbol or a metaphor for life, it may be helpful to have them compare the map to other forms of self-portrait. One common thread shared by the map, the photograph, the home movie, the blog, and the memoir is that they depict life in miniature—they abbreviate and condense an entirety of experiences into something that is compact, portable, and easy to convey. Emphasizing this might allow students to understand the potentially alien concept of rendering life as a series of towns, mountains, forests, and oceans.

In order to aid students in "reading" Atwan's map, consider exploring it as a group, asking them to name the real-life equivalents of the various features. After establishing that the map is in many ways based on the real world (mainly the United States), ask students where the map takes leave of reality, how it is different from a conventional map of a real place. Allow them, if necessary, to look at conventional maps for comparison. Have them also take note of Peter Cross's cartoonish illustrations—the disproportionately sized icons, the dragon, Neptune, the mermaid—and ask them how these elements contribute to the overall tone of the map.

MESSAGE

Bobland raises questions about colonization by poking fun at it. The oversized banner and fancy logo are meant to remind us of flags planted by famous explorers centuries ago—"I proudly claim this land in the name of Bob." The name, however, has a comical ring, as does the incorporation of "Bob" into literally every feature on the map. *Bobland*, where all roads (and rivers and forests) lead to Bob, reminds us of actual places named after people—or their respective religions or sovereigns—who claimed the glory of reaching lands previously unknown to them.

METHOD

Students should immediately note that the country Bobland is reminiscent of the United States, with corresponding plays on the names of its cities, bodies of water, and noteworthy geographical features. Despite the fact that the name of every feature on the map is some derivative of "Bob," we are meant to understand the real-world equivalents through placement as well as graphical cues (Bobbywood is Hollywood, Las Bob is Las Vegas, and so on).

MEDIUM

Bobland is drawn in the style of maps from Western Europe's Age of Exploration, fifteenth- and sixteenth-century works of art inspired by the voyages of such figures as Christopher Columbus, Marco Polo, and the Spanish conquistadors. The cartographic style is meant to emphasize the idea that Bobland is previously undiscovered terrain, and so it is well suited to the task of self-portraiture.

ADDITIONAL WRITING TOPICS

1. A map isn't as immediately obvious a choice of medium for constructing a self-portrait as is a painting or a reflective essay. Spend a few minutes brainstorming other unconventional genres and media for this task, choose one, and use it to construct your own self-portrait.
2. Based on the details in *Bobland*, what can you deduce about the type of life Robert Atwan means to portray? What can you tell about his actual life, as well as the type of life he imagines for himself? Write a short narrative that highlights some of the key points in Atwan's life, using the map as a guide.
3. Atwan explains part of the reason he made his map: "It has been my fate never to live anywhere I really liked." *Bobland* is his attempt to fix that problem by resorting to imagination in order to augment his reality. If you could live anywhere in the world (or beyond), where would it be? Write a description of this place, explaining how you foresee your life unfolding based on this choice.

DIVERGENCES

Print

Abbott, Edwin A. *Flatland: A Romance of Many Dimensions*. 1884; reprint, New York: Dover Thrift Editions, 1992. This novella about inhabitants of a two-dimensional world called Flatland uses maps and illustrations to explain concepts of multiple dimensions—and it's an allegory that satirizes the social situation in Victorian England to boot.

The New Yorker, *Troika*, *The Baffler*. These three magazines are similar in content and humor to the now-defunct *Wigwam,* in which *Bobland* appeared. Sifting through back issues of magazines such as these could invite discussion on what constitutes humor and how the criteria change depending on the makeup of the particular audience involved.

Web

http://www.ihrinfo.ac.uk/maps/. "The Map History/History of Cartography" site sponsored by London's Institute of Historical Research is a very thorough online resource that contains several digital facsimiles of old maps that students can peruse for purposes of comparison.

FAMILY PHOTOS
Sally Mann

PHOTOGRAPH, INTERVIEW

In her interview with Melissa Harris, Jessie Mann mentions the most controversial aspect of her mother Sally Mann's work: "I . . . think she [Sally Mann] brought out a certain sexuality in children that nobody wants to think about. I think if you have a certain background or beliefs those photographs could be upsetting or offensive. I don't agree with that point of view, but maybe there's something to their idea that that part of children shouldn't be played up." Even though it has been nearly a decade since

Mann's *Immediate Family* was published, some people still find the work controversial. Also, the recent surge in popularity of teen pop sensations such as Britney Spears and Christina Aguilera suggests a cultural shift in the attitudes to child and adolescent sexuality; many teen and preteen fans of these singers mimic their occasionally provocative mannerisms and style of dress. Since many of your students are probably familiar with these celebrities, this comparison should spur a debate concerning whether or not Mann's photography is intended to present the sexual nature of childhood.

You may want to direct students to the dual roles occupied by Sally and Jessie Mann. In one sense they are mother and daughter, and in another they are photographer and model. Students could examine Jessie's interview with these multiple roles in mind, and they could even point out places in the text where she speaks as a daughter (where she admits feeling defensive and protective when her mother is criticized) and as a model (where she discusses the artistic merit of Mann's work). You might suggest that they focus on the shifts in tone and diction in Jessie's speech, specifically those places where she refers to Sally as "Mom" or when she uses the more distant title "artist." What do these differences reflect in terms of Jessie's argumentative strategy as she tries to explain or defend Sally Mann's art?

MESSAGE

The photo challenges our cultural concepts about childhood ("candy") and adulthood ("cigarette"). Mann's point seems to be that we deny that certain aspects of our children's identities actually exist. An accompanying caption to the photo could read: "By her dress, stature, and features, we can identify her as a child, but her facial expression, her posture, and the accomplished way in which she holds her cigarette suggest a girl on the verge of growing up."

METHOD

Though the cropping is casual—the figure on the right has seemingly been pushed to the edge of the frame for no clear compositional reason—Jessie's stance suggests that she is posing; note especially the direct eye contact she makes with the camera. This ambivalence creates a kind of generic dynamic ("Is it casual? Is it contrived?") that sustains the viewer's interest, raising the question of whether Jessie's pose is her own or one suggested by her mother.

MEDIUM

Photography, like other visual arts, does not convey information on a verbal level, so when this level is introduced, our perception of the art changes. Some students may think that Jessie Mann's interview augments the photograph, while others think it detracts from it. Both the interview and the photograph are different takes on Jessie Mann, and it is difficult to say which medium is more direct, either concerning Jessie Mann or her relationship with her mother. However, the interview does produce a portrait of a woman—Sally Mann—who struggled to balance the roles of artist and mother.

ADDITIONAL WRITING TOPICS

1. Critic Julian Fell wrote of Sally Mann's parenting, "I'll say that seems a rotten way to bring them up." Do you agree or disagree with that assessment? How does reading the interview with one of Mann's three children influence your attitude? Write an essay that either supports or rebuts Fell's comment.

2. Sally Mann has granted several interviews, but the one presented in this text is with the daughter rather than the artist. Why do you think this selection was made? How might your reaction to *Candy Cigarette* be different if an interview with Sally Mann were included instead?
3. Working with a partner, imagine a scenario in which you each feel compelled to write a letter to the editor of your hometown newspaper regarding a recent Sally Mann exhibit at your local museum. One of you should construct the argument that Mann's art is offensive and needs to be taken down. The other should defend Mann's art and its right to stay in the museum. Collaborate while drafting the letters to better understand both sides.

DIVERGENCES

Print

Sally Mann. *At Twelve: Portraits of Young Women*. New York: Aperture, 1988.

———. *Immediate Family*. New York: Aperture, 1992.

Web

http://www.britneyspears.com. The official Britney Spears Web site, with several pictures.

http://www.dazereader.com/sallymann.htm. Daze Reader's site features a very thorough collection of online Sally Mann resources, including a profile, links to several articles, galleries, and image archives.

Audio/Visual

Kids. Directed by Larry Clark. 91 minutes, rated R. Trimark Studios, 1995. Videocassette. Clark's controversial faux documentary explores the sordid underground of New York City youths, with several unsettling depictions of unknown young actors in adult situations.

TABLOID PHOTOGRAPHY
Weegee

ESSAY, 4 PHOTOGRAPHS

Weegee's photography might be considered the exact opposite of the work of Sally Mann or Cindy Sherman. Instead of staging a photograph so as to make it appear candid, Weegee depicted real-life events in such a way that they appeared artfully composed: for Weegee, life imitates art. You may want to discuss with students what sort of statement Weegee tried to convey regarding this relationship in his photography.

Wendy Lesser's essay gives students an excellent model of how to read a photograph. It might prove valuable to have students analyze Lesser's structure so that they better understand their own goals when reading visual texts. They should notice the kinds of elements to which Lesser draws her readers' attention, such as the positioning of the figures, the seemingly insignificant props, and the background composition, as well as the kinds of conclusions she draws about the event itself and the feelings of the moment that Weegee has captured.

Finally, students should focus on the photographs themselves. Remind them that the images were not intended as art when they were first shot, but as photojournalism. Ask them to look at the pictures with both styles in mind. Also ask them what sort of news stories Weegee's photos might have accompanied and have them compare the quality of his work to that of tabloid photojournalism as it stands today.

CALLOUT QUESTION

Weegee composed his photograph based on the frame suggested by the event itself—the frame of the car door allows Weegee to focus on the interaction of the two people inside, instead of the damage done to the vehicles. The incident itself is understated, though the title and the boy's bloody face give us clues about what has happened outside the photograph. The reflection amplifies the event, emphasizing the boy's condition above all other elements in the accident. This selective framing, as well as the relatively close distance, implies a certain compassion on Weegee's part to convey the intimate and vulnerable mood of the moment.

MESSAGE

Because *Newsboy* appears so deliberately composed (even the paper's headline is not obscured by the boy's hand), it is difficult to think of it as a casual snapshot. Lesser calls it a self-referential photo, a vehicle for the same tabloid journalism that Weegee produces. Though the event of Black Dahlia's confession is not trivial, it does tend to take a back seat to another message in the picture—namely, that we are obsessed with sensational news. Without the newspaper, the picture of the boy would be unremarkable and ordinary, and the social commentary would be lost.

METHOD

Dancing and *Mulberry Street Café* prominently feature audiences that serve as foils to the actions of the main characters and give the viewer hints on how to react to the photograph. The viewer is associated with the audience because both positions serve similar functions—looking at the main action without necessarily influencing it and taking pleasure in witnessing without being involved. The onlookers' warm smiles in *Dancing*, for instance, indicate to viewers that they should feel sentimental about the main action in the frame; the girl in *Café*, however, shares with the viewer a disapproving expression (see below).

MEDIUM

Students may want to consider the divide between the textual and the visual in this photograph's composition—why did Weegee "waste" so much space by including the café's sign? By far, though, the dominant elements in this photo are the human subjects, particularly the three figures on the right and the little girl on the left. The men in the café all seem energetic and talkative, and they are involved in their respective activities to the point of ignoring the camera entirely. As a result, the viewer is drawn to the little girl, who is positioned outside the action in this photo and, though she doesn't look at the camera, faces it in such a way that suggests she is aware of its presence. We tend to identify with her as spectators, and may even take her expression and posture as indications that she isn't exactly impressed with the goings-on inside.

ADDITIONAL WRITING TOPICS

1. Examine the contents of a current tabloid such as the *National Enquirer* or *Weekly World News*. What similarities or trends do you note among the articles? How would you characterize the audience for this type of news? Write a mock article for one of these newspapers, emulating its style.

2. Look closely at Weegee's four photographs. What do you take to be the "New York character" of these pictures? Describe the types of photographs you would take that would likewise characterize or represent the town where you live.
3. Choose one of the photos Wendy Lesser discusses in her essay. Given her description of the picture you've chosen, write a short narrative based on the picture.

DIVERGENCES

Print

Gamson, Joshua. *Freaks Talk Back: Tabloid Talk Shows and Sexual Nonconformity*. Chicago: University of Chicago Press, 1998. Gamson critiques television's talk-show genre, providing some historical background on its newspaper counterpart.

Weegee. *Naked City*. New York: Essential Books, 1945.

Web

http://www.icp.org/weegee/. The International Center of Photography's well-designed Weegee site offers several images, biographical information, and several audio clips from interviews with the photographer.

Cinderella
William Wegman

PHOTOGRAPH

W. C. Fields is noted as saying that in the world of show business the worst people to work with are animals and kids; William Wegman's patient weimaraners seemingly dispel half of that rule. Wegman's photography makes use of a technique called anthropomorphism, or the attribution of human qualities to something that isn't human. Note that in works like *Cinderella*, Wegman's point is to stress how silly, how unnatural, living in a human culture actually is; when his dogs stand in place of us, the images emphasize the constructed nature of wearing high heels and lingerie, donning a suit and tie for work, and even getting married. You might also want to point out that Wegman is not the only person to use this technique. In fact, human culture has a long tradition—in paintings, films, and folklore (think back to Aesop's fables)—of giving animals the ability to speak and act like people. Discuss with your students why we have this urge to make animals resemble us.

Another important consideration of Wegman's art could be the duration of this particular photography project—Wegman has been posing his weimaraners ever since the late 1970s, and what started out as a set of shrewd statements on the conventional and scripted elements of our lives and culture may no longer offer us the same insightful critique it once did. With the advent of merchandising in the art world, can Wegman's art still carry its critical impact? Broadly, this question invites discussion about the relationship between art and commercialism. What does the fact that one can purchase, say, a coffee mug decorated with a reproduction of Van Gogh's *Starry Night*, an inflatable version of the figure in Edvard Munch's *The Scream*, or a T-shirt with one of Wegman's dog photos say about our culture's attitude toward art, not to mention the ability of art to continue providing relevant social critique?

MESSAGE

With its idyllic outdoor background and beautifully elaborate costuming, Wegman's photo is in all ways true to the Cinderella story, but it is the presence of the dogs themselves that evokes surprise. Their humanlike positions and expressions almost seem to be from the world of animated cartoons, but the fact that this staged portrait happened in "real life" is the source of our amusement.

METHOD

Supplying students with pictures of other dog breeds such as the German shepherd or pit bull will certainly invite productive comparisons to the weimaraner. The weimaraner has a gentle and calm aspect that makes it a good model for the type of photography Wegman orchestrates. (Would a pit bull look winsome in a wedding dress?) Additionally, their unchanging expression, suggestive of an aloof and unimpressed demeanor, makes the pictures so much more funny.

MEDIUM

Cinderella is in every way appropriate to the wedding-portrait genre—it is romantic, dignified, and holds a certain reverence—except for the models. By playing the role of the wedding photographer, the cheerleader in the background who tries to get his subjects to express their unfaltering love for one another (and in this case fails), Wegman seems to be providing some satirical commentary on the romantic, fairy-tale myth we often associate with marriage and love.

ADDITIONAL WRITING TOPICS

1. Critics of Wegman have suggested that his dog photographs and videos are not serious art but rather sappy imagery that plays on the audience's sense of pathos. (Wegman's work has appeared on several calendars, commercials, and even *Sesame Street*.) Do you agree or disagree with this assessment of Wegman's work, or do you think it makes a sophisticated artistic statement? Defend your choice.
2. Spend some time looking at a few of Wegman's other photographs besides *Cinderella*. Based on the various poses and themes that Wegman uses, what type of relationship do you imagine he has with his weimaraners? Write a brief but detailed narrative of what you imagine his life is like with his pets away from the camera.
3. Brainstorm and then write a list of settings and themes that would be appropriate for Wegman's work. Next, construct a list of scenarios that might not make good choices for a typical Wegman photograph. By looking at these two lists, can you also describe the qualities, themes, and ideas that are characteristic of Wegman's photographs?

DIVERGENCES

Print

Kunz, Martin, ed. *William Wegman: Paintings, Drawings, Photographs, Videotapes*. New York: Abrams, 1990. This volume contains several essays about Wegman's work in different media as well as an interview with the artist conducted by David Ross.

Wegman, William. *Cinderella*. New York: Abrams, 1993.
———. *Fashion Photographs*. New York: Abrams, 1999.
———. *Little Red Riding Hood*. New York: Abrams, 1993.
———. *Man's Best Friend*. New York: Abrams, 1982.

Web

http://www.annegeddes.com/. Anne Geddes does with babies what her contemporary William Wegman does with weimaraners. This site, which is slightly more commercial than Wegman's, could be used in an interesting comparison/contrast exercise.

http://www.wegmanworld.com/. Wegman's own copyrighted site includes samples from his portfolio, background on his dog models, and merchandise.

REFLECTIONS
Norman Rockwell

2 PAINTINGS, PHOTOGRAPH

You might begin looking at the pieces in this selection by asking students to talk or freewrite about what comes to mind when they hear the term "Americana." What sort of images, phrases, time frame, and other ideas do they associate with the term? Particularly, how are these associations brought forth in the paintings and illustrations of Norman Rockwell, identified by many as the paragon of Americana art? Generally, you could ask students to make distinctions between the types of art that merely idealize our popular conception of America and those that are more explicitly identified as propaganda.

Norman Rockwell is generally regarded as a talented and studied artist by fans and critics alike, but there is disagreement about the function of his particular brand of realism—is it merely naive and clichéd sentimentality, or is it a subtle attempt at complex symbolism? The comments by Charles Rosen and Henri Zerner in this selection offer especially good readings of the way Rockwell used props in his paintings for symbolic effect. As Rosen and Zerner point out in several places, some of Rockwell's props serve dual purposes, depending on who he imagined his audience to be. For example, the eagle atop the mirror frame and the Parisian fireman's helmet in *Triple Self-Portrait* might serve as simple ornamental touches for his popular audience, but they also function as symbolic commentary to knowing avant-garde artists. Rhetorically, this is an interesting and economical technique that students should notice and understand.

Finally, it might prove an interesting exercise to have students look at a sampling of magazines from Norman Rockwell's heyday (particularly those he illustrated) to today. Have them take note of the changes that have occurred over time. How has the ratio of text to graphics changed? What types of graphics are used now? What shifts in visual layout and style can they identify? How has the editorial content changed? What conclusions can they draw about the cultural attitudes of America fifty years ago versus today?

MESSAGE

Obviously, Rockwell's self-portrait contains three main images of himself as well as several small sketches pasted to his canvas. You might press students to consider the concept of portrait more broadly in Rockwell's painting, for as Rosen and Zerner indicate, many of the props in the painting carry symbolic meaning important to Rockwell's identity. The ceremonial fireman's helmet (the Parisian avant-garde's symbol for schooled painters) shows that he considers himself an academic artist; the eagle atop the mirror's frame shows his understanding of himself as a purveyor of Americana; the reproductions of self-portraits by Van Gogh, Rembrandt, and others perhaps show his desire to achieve the talent of these Old Masters. Finally, by titling the work *Triple Self-Portrait*, Rockwell is acknowledging the self-reflective nature of portraiture and of art in general.

METHOD

As Rosen and Zerner note regarding *Girl at Mirror*, Rockwell has lengthened the slip, changed the girl's expression, and included props such as a broken doll, a tube of lipstick, and a picture of Jane Russell. These additions provide a context for the image that suggests a girl who is playing dress-up. These changes upset the critics, in part because the story conveyed in the Rockwell version was simply too heavy-handed and sentimental. They much preferred the photograph because of its subtlety and ambiguity.

MEDIUM

The styles of *Triple Self-Portrait* and *Girl at Mirror* are similar, particularly in Rockwell's use of thin, wavy lines and his centrally balanced sense of composition. Some themes appear in both paintings as well; ask students to consider the role that mirrors play in each painting. What sort of metaphor is Rockwell constructing? Applying Rosen and Zerner's critique of *Girl at Mirror* to *Triple Self-Portrait*, ask students to describe how the painting would change if the symbolic props were taken out—the helmet, the mirror's elaborate frame, and the reproductions of the Old Masters' self-portraits.

ADDITIONAL WRITING TOPICS

1. After viewing several of Rockwell's *Saturday Evening Post* covers in class, evaluate the tone of the paintings. What elements do they share, and what sorts of feelings do you think the artist is trying to convey through his use of shading, props, and overall composition? How well do Rockwell's covers illustrate the contents of the magazine?
2. Interview someone you know who grew up during the 1950s, and ask him or her to describe the cultural climate of the time—the major political events, attitudes toward spirituality, feelings about America, race relations. How well do Rockwell's paintings match this description? If there are any distinct differences, how do you explain them?
3. Research some American fiction and poetry from the mid-twentieth century. Find a piece of writing that you think nicely parallels Rockwell's

paintings in terms of tone, imagery, and content. In an essay, citing quotes from your choice, explain why your selected writing is "Rockwellian" in character.

DIVERGENCES

Print

Frost, Robert. *Complete Poems of Robert Frost*. New York: Holt, Rinehart and Winston, 1964. Critics have often condemned Frost's poetry in similar fashion to Rockwell's illustrations—as artless sentimentalism. Also like Rockwell, Frost has undergone a revival, with advocates suggesting that his writing is more dark and complex than he has been given credit for.

Spiegelman, Art. *Maus: A Survivor's Tale*. New York: Pantheon Books, 1991. Spiegelman's Pulitzer Prize–winning graphic-art novel recounts his family's struggles during the Holocaust. Spiegelman considered Rockwell an influence, and sees his own work as a satirical extension of Rockwell's less explicit symbolism.

Web

http://www.jokeindex.com/joke.asp?Joke=2795. Jokeindex.com's parody list of the thirty-one worst Rockwell painting titles could prompt a discussion of why the parody titles are funny, and what values they assume for Rockwell's art.

http://www.nrm.org/. The Norman Rockwell Museum at Stockbridge online site contains, among much else, an online tour of the museum.

http://www.paonline.com/zaikoski/rockwell.htm. This page of Pennsylvania Online offers several high-resolution images of *Post* covers.

Audio/Visual

Crumb. Directed by Terry Zwigoff. 119 minutes, rated R. Columbia/Tristar, 1995. Videocassette. This is the acclaimed documentary of the controversial and often tasteless comic book artist R. Crumb, whose antiestablishment and anti-family-values satire makes for an interesting counterpoint to Rockwell's art.

Physicians Against Land Mines

ADVERTISEMENT

If your students are not very familiar with the issue of land mines, some further research may be in order so that they better understand the Physicians Against Land Mines advertisement. A visit to PALM's Web site, as well as to sites of other activist organizations concerned with the problem of land-mine deactivation, would provide necessary background for discussing the argument made in the ad. Discuss with your students the validity of the medium, a magazine advertisement, for getting political messages to the public. PALM's ad, in terms of its design, location, and overall aesthetic, conforms to our ideas about commercial advertisements—does this serve to "cheapen" the message, or is it simply an effective means of reaching a mass audience?

This ad would serve as a good text for discussing the various rhetorical strategies involved in constructing an argument. Specifically, you could point out the ad's different appeals, as well as the degree to which each type of appeal is made. For instance, PALM's ethical position is conveyed in its name as well as in the tone of the ad copy—the organization's members identify themselves as physicians, without any overtly self-serving political agendas, and the compassionate tone of the message paints the organization as sincere and trustworthy. Also, the text of the ad appeals to an audience's sense of reason by offering statistics as evidence and using a tone that suggests a logical course of action. Finally, and most prominently, the photo of Emina Uzicanin appeals to the reader's emotional sensibilities. The ad's designers knew that the relatively large size of the image, as well as the fact that Emina makes eye contact with us, would have a powerful impact.

CALLOUT QUESTION

Though the use of "its" might well be unintentional, the closest antecedent would be "planet." This may be a subtle rhetorical strategy, one that suggests the unity involved in the campaign. To call Emina a victim of the planet makes all of us in some way responsible to her and those like her, and the urgency of helping takes on a magnitude far greater than ethnicity or the boundaries of foreign countries; it defines everyone as part of the same community.

MESSAGE

The advertisement tries to unsettle us by presenting the image of a girl, one who is otherwise normal looking except for the missing leg. At first glance, the amputated leg is not obvious because of the ad copy's placement, and the reader does a double take on noticing the missing limb. The fact that a young person is used in the ad is important; because she is dressed in typical clothes (T-shirt and shorts) and has no horrible facial deformities, readers can better identify with her and empathize with the ad's plea for support.

METHOD

Again, aside from the missing leg, the attempt is to portray Emina as much like us as possible. Rhetorically, PALM wants to stress the message that the land-mine problem is one that affects everyone, and to these ends they erase as many specifics as possible. To offer the reader visual cues that indicate a foreign location would distance the reader from the problem, as would showing Emina with prosthetics or crutches. The ad's use of the bare branch is symbolic of death or of youth's untimely demise, the oftentimes literal consequences of land-mine accidents.

MEDIUM

The text of the ad causes the reader to focus attention on Emina's missing leg—our eyes have to linger on the space where her leg should be in order to read the information, which reinforces that message. In the consistent effort to present the reader with as few specific details as possible (so that they are more likely to form associations with the message), the Leo Burnett firm gives us very general statistics about the global nature of the land-mine problem: the total rate of land-mine accidents, the total number of active mines, the number of countries where this is a problem. Consequently, we are given no specific details about Emina Uzicanin's situation; she serves as a symbol of a larger problem in the ad. As indicated in the headnote, this ad is more likely to appear in magazines with liberal political leanings (*Bomb*, *Atlantic Monthly*, *Harper's*) than in conservative publications (*National Review*) or those with no overt political agendas (*Jane*, *Entertainment Weekly*).

ADDITIONAL WRITING TOPICS

1. As the headnote states, the PALM campaign has been supported by such magazines as *Harper's*, *Atlantic Monthly*, *Bomb*, and *People*. Find examples of these magazines and look through their contents. Analyze and describe the typical audience for each—who is likely to read them, and who is not?
2. Have you ever contemplated the possibility of losing one of your limbs in an accident? How do you think you would adjust to this condition?
3. The PALM advertisement was designed by Leo Burnett, a commercial advertising company. Does knowing this affect how you read the message conveyed by the ad? Is commercial advertising an effective or appropriate forum for political activism? Why or why not?

DIVERGENCES

Print

Heller, Steven. *Design Literacy (Continued): Understanding Graphic Design*. New York: Allworth Press, 1999. This series of essays focuses on important graphic design works of the past century, many of which comment on graphic art's capacity for political statement.

Web

http://www.banmines.org. PALM's official site offers several case studies, statistics, video clips, and graphic images related to devastation caused by abandoned land mines.

http://www.care.org. This site for the global relief organization Cooperative for Assistance and Relief Everywhere (CARE) offers additional information on demining initiatives and has an archive of public service announcements about humanitarian land-mine projects that could be compared to PALM's ads.

Audio/Visual

A Plague of Plastic Soldiers: Land Mines in Cambodia. Produced by Stephen Smith. Soundprint Media Center (1-888-38-TAPES), 1997. Audiocassette. Minnesota Public Radio reports the tragic aftereffects of active land mines left in the Cambodian countryside.

2 Telling Secrets

Since the selections in this chapter all focus on the nature of secrets, and the relationship between public and private life, it is important to remember that at the center of this topic lies the idea of truth. Learning a secret is really about learning the "truth," and keeping a secret demands that the "truth" remain hidden. When we learn about the "real life" of a famous actor or politician, we are fed the illusion that we are in fact learning about what that person is "really like," that is, the truth about his or her private life.

The mainstream media thrive by feeding the public "true" stories, by telling the secrets of famous people's private lives. You should begin this chapter by asking students what secrets they've learned about a famous Hollywood actor recently or what is on the cover of this week's *People*. This should lead to a discussion about secrets and secret-sharing: Why are humans infinitely interested in rumors, gossip, and secrets concerning people we will most likely never meet and facts that barely affect us? This fascination with hidden truth appears on all levels, from hushed tones among families to high-level, classified government documents. It appears that our society functions in part because of secrets and the truth behind them. The omnipresence of secrets is an important notion to reiterate to your students throughout this chapter. Remind students as often as possible of the role secrets play in their own lives. By making a personal connection with the material, students should fill their writing with secrets, both hidden and uncovered. Ask students both to be personal and to refer to celebrities, politicians, or other famous figures—who will surely provide endless writing fodder.

SELLING SECRETS
Miss Clairol

ADVERTISEMENT, ESSAY

You can begin this "Does she . . . or doesn't she?" selection by considering Shirley Polykoff's advertisement in its historical context—have students discuss (and research, if necessary) women's social condition in the 1950s. Direct them to focus on the kinds of political or employment opportunities available to women as well as their general social status and the sexual politics of the time. In light of this background, students may more clearly see the impact of the ad's somewhat risqué message, which may have been lost on them before. An interesting follow-up exercise might be to have students brainstorm a list of messages and images that are considered taboo or shocking for us today.

You could also spend some time teaching students how to "read" the ad itself. Have them consider the layout of the page and the ratio of image to text. You might ask them to rank the individual elements in the ad—such as the photos, the headline, the ad copy, and the Good Housekeeping Seal—in order of importance. Also, direct discussion to address how the product itself is located within the ad (a small picture in the right-hand corner)—is the point of this ad to sell hair coloring or a new image?

Aside from its many astute observations about the logic used in advertising in general (e.g., ambiguity, making impractical items appear common), James Twitchell's essay offers students a great example of how to analyze an ad. The method Twitchell uses is a combination of historical research—documentary evidence surrounding Polykoff's development of the campaign—and rhetorical analysis that considers the anticipated receptions of the message and the audience's actual reactions. After recognizing this structure, students could apply or adapt this method to their own analyses of contemporary advertising culture, from Abercrombie & Fitch catalogs, to Volkswagen commercials, to Promise Keepers billboards, to Target branding.

MESSAGE

Students should understand that for the comparatively conservative social climate of the 1950s, Polykoff's message could have been deemed too racy for the public because of its sexually suggestive tone. While women may have intuitively understood the phrase to be about their hair color, for men the phrase was reminiscent of the type of hushed conversation you might overhear in locker rooms or in a huddle around the office water cooler—wondering about a young lady's sexual potential.

METHOD

As suggested above, the Miss Clairol campaign derived ambiguity from the slogan, which men and women readers would take differently. Ambiguity, as noted in the chapter's introduction, is a successful technique because it plays with an audience's delight in encountering surprise and secrecy—simply put, this strategy sells the ad itself, and hopefully the product as well.

MEDIUM

The Miss Clairol campaign featured mothers in part because they were seen as older women and therefore among the demographic group most likely to use the product. At the time, hair color was predominantly used to cover graying hair, and there was still a stigma associated with young women changing their hair color. The woman in this particular ad is depicted only with a young child because the ad's message plays on the element of secrecy (the husband shouldn't know about his wife's hair—only the hairdresser). Clairol ads of today have abandoned these strategies, opting for more liberated, individualistic slogans and youthful models—students should have fun comparing the two sets of ads, whose differences are rather striking.

ADDITIONAL WRITING TOPICS

1. Locate a variety of hair-color advertisements dated since the Miss Clairol campaign, which ran from the mid-1950s to the early 1970s. In what ways have the selling strategies changed over the years? What does this say about our society's change in attitude about people coloring their hair?
2. In the first paragraph of "How to Advertise a Dangerous Product," James Twitchell outlines two types of commercial products—the common and the radical—and two respective strategies used by advertisers to market them—making the common appear "new and improved" and making the radical seem acceptable and commonplace. Do you find this theory true or complete? Either support Twitchell's claims by analyzing a few advertisements with his outline in mind, or refute his claims by locating advertisements that don't match his theory and discussing how they do not.
3. As the Miss Clairol selection indicates, ambiguity is often used as an effective technique by advertisers and marketing professionals. Choose three or four contemporary advertisements (from any medium) that use this technique and, in an analytical essay, tell how the message works. What conflicting ideas does the ad present, and how does this conflict work to appeal to potentially different types of customers?

DIVERGENCES

Print

Twitchell, James B. *Adcult USA: The Triumph of Advertising in American Culture*. New York: Columbia University Press, 1996.

———. *Twenty Ads That Shook the World: The Century's Most Groundbreaking Advertising and How It Changed Us All*. New York: Crown Publishers, 2000. This collection includes P. T. Barnum's cons, Coke's Christmas ad, the Marlboro Man series, and Apple's 1984 Superbowl commercial.

Weiss, Jessica. *To Have and to Hold: Marriage, the Baby Boom, and Social Change*. Chicago: University of Chicago Press, 2000.

Web

http://www.artshum.org/pages/dream.html. Jan Kurtz's online essay "Dream Girls: Women in Advertising" outlines the historical shifts in feminine ideals from the 1950s onward.

http://www.chickenhead.com/truth/1950s.html. The Truth in Advertising page on the Chickenhead humor/satire site houses an image gallery of cigarette advertisements dating from the 1950s.

Audio/Visual

Beyond Killing Us Softly: The Strength to Resist. Directed by Margaret Lazarus. 33 minutes, not rated. Cambridge Documentary Films, 2000. Videocassette. This documentary offers an analysis of the representation of women in print and television advertising as well as in new media. Featured are such noteworthy feminist thinkers as professor and author Gail Dines; psychologists Carol Gilligan, Catherine Steiner

Adair, and Valerie Batts; activist Gloria Steinem; and third-wave feminist and cybercolumnist Amy Richards. For ordering information, contact the distributor at www.cambridgedocumentaryfilms.org or 617-484-3993.

What Secrets Tell
Luc Sante

ESSAY

One immediately apparent signature element of Luc Sante's writing is his use of vivid figurative language. You could start a discussion of "What Secrets Tell" by having students compile a list of some of the striking metaphors, similes, and descriptive details they find in it (a sure task). Examples include a series of similes to secrets ("black holes," "land-speed records," "lotteries"); the eyebrow-raising examples of Romantic Secrets; and the comparison of the Internet to Borges's library of Babel. After noting some examples, students could explain why they think the author might have chosen a particular word or phrase over something more conventional or trite. They can then elaborate on how these choices contribute to the author's overall style and attitude toward our society's current cult of secrecy.

As a good model of an expository essay, "What Secrets Tell" can be discussed in terms of its structure, especially in its attempt to classify the various types of secrets in American culture, including the personal, the state, and the atomic levels. As the exercises in the text suggest, one way to deal with this aspect would be to have students offer alternative systems of classification. Another exercise might involve having students come up with specific examples of Sante's types.

This essay shows how the phenomenon of public confession pervades every sphere of our culture. Sante argues that all aspects of our lives are subject to the temptation to reveal our secrets, and this temptation is made stronger by the media that continually help make such revelations public. You might finally ask your students to identify the different features or characteristics of television, newspapers, radio, computers, or other media that construct or facilitate these cultural pressures.

MESSAGE

Part of Sante's main argument is that secrecy and disclosure are both part of the human condition; the phenomenon reaches across cultures and is practically global. According to Sante, people need to establish an unknown in this existence because, otherwise, the only unknown factor of existence would be death. Sante's last sentence sums this up: "Secrets hold out the promise, false but necessary, that death will be deferred until their unveiling."

METHOD

Sante's classification system works on a principle of scale—he starts with the smallest, most individualized types of secrets and then expands to those types that might matter to communities, nations, and finally the world and beyond. Although there is a logical structure to the classes, students might notice the somewhat arbitrary assignment of secrets to categories, as well as outright omissions (Where, for instance, are secret crimes?). Students can experiment with this grouping technique by considering other logical structures such as chronology (How do secrets change over historical time or over a person's lifetime?) or geography (Are there different types of secrets in rural versus urban settings?).

MEDIUM

"What Secrets Tell" refers to so many visually oriented texts (*The Jerry Springer Show*, the Internet, the Watergate hearings) that it could easily lend itself to a documentary treatment. Producing a short documentary might make for an engaging, though ambitious, group student project, invaluable for practicing a variety of skills from rhetorical strategy, to artistic adaptation, to video/digital editing.

ADDITIONAL WRITING TOPICS

1. Sante writes that while the memoir has declined in popularity, confessional television talk shows are going strong after a decade of existence. In his mind, this is because "television thrives on repetition." What does Sante mean by this? Do you agree or disagree that this is the reason for the talk-show genre's popularity? Explain the reasons behind your answer.
2. Sante writes, "Anybody who doesn't carry around one or two secrets probably has all the depth of a place mat." Think of some secrets you have carried. Which of these have you disclosed? Did you confess to someone close in your life or to a complete stranger? Recount one such event in a short narrative, and be sure to include the consequences of your confession.
3. Sante offers several different categories of secrets. Alone or in small groups, compose hypothetical examples to fit in these categories. Try to be as detailed as possible in your descriptions so that it is clear why you placed each secret in each particular category.

DIVERGENCES

Print

Sante, Luc. *Evidence*. New York: Farrar, Straus, and Giroux, 1992.

———. *Factory of Facts*. New York: Pantheon Books, 1998.

———. *Low Life: Lures and Snares of Old New York*. New York: Farrar, Straus, and Giroux, 1991.

Wicks, Robert. *Understanding Audiences: Learning to Use the Media Constructively*. Mahwah, N.J.: Lawrence Erlbaum, 2001.

Web

http://members.tripod.com/mking60/conspiracies.htm. This personal Web site compiles several conspiracy theories and global threats, the "atomic secrets" Sante's essay mentions.

http://www.npr.org/programs/death/readings/stories/sante.html. NPR's Web site has text and an audio file of Sante reading an excerpt from "The Unknown Soldier."

Audio/Visual

The Ringmaster. Directed by Neil Abramson. Starring Jerry Springer. 95 minutes, rated R. Artisan, 1998. This behind-the-scenes spoof of *The Jerry Springer Show* deals with the secret lives of the show's guests, some of whom fabricate their stories in order to get fifteen minutes of fame.

HAUNTED
Empty Houses

PAINTING, ESSAY

Joyce Carol Oates begins her essay "They All Just Went Away" by admitting that she isn't sure why she is fascinated by abandoned houses. She then proceeds to explain to us this fascination by drawing on her childhood memories, some painful, some shameful, some wonderful. Begin discussing this selection by thinking about the different purposes and goals that go into writing in general. Students could look at this piece as a kind of therapy—Oates writes in a way that benefits both herself and her readers.

Oates's essay presents students with a collection of fully developed themes: the connection between femininity and domestic space; the pain associated with the loss of childhood innocence and the development of adolescence; the association of the lower class with shameful secrets; and the role of evolution in the development of women, families, and humanity in general. The essay could also offer an excellent opportunity to illustrate to students how a theme is sustained throughout a piece of writing. As an exercise, consider having them construct a list of themes, dividing them up among small groups, and glossing the essay for places where a particular thematic development occurs. Students could then work on similar strategies in their own writing, employing metaphors, sustained analogies, and other rhetorical devices.

When dealing specifically with Edward Hopper's *Haunted House*, try asking students what about the painting indicates that the house is haunted. Does Hopper use any special techniques, such as a particular type of brush stroke, certain color combinations, use of shadows, or a stylized depiction of the sky and background? One interesting group project would be to have students represent a haunted house in a medium of their own choosing—painting, short story, model, computer graphics. Have them follow up this project by writing a short rationale for the different choices they made in constructing their houses.

While Oates explicitly compares the painters Edward Hopper, Charles Burchfield, and Wolf Kahn to each other, she implicitly makes a comparison between painting and her own artistic medium. Students could brainstorm on the various ways in which painting and writing differ, as well as on the strengths and weaknesses of each medium. Of Hopper specifically, Oates writes that he is a "surrealist"; she describes his technique as emoting "a kind of rage, a revenge against such restraints, in Hopper's studied, endlessly repeated *simplicity*." After showing them examples of paintings by the other artists Oates mentions, discuss with your students how well Oates has judged the aesthetic quality of these works.

MESSAGE

Oates's attraction to abandoned houses may have simply been the result of a child's curiosity, but it also came from a young woman's desire to discover herself. In part, the notion of abandonment parallels the leaving of the womb at birth, and Oates marvels at the excitement of trying to trespass on this forbidden, "maternal" space. The term "abandoned" also can be applied to Oates's view of the Weidels, who abandoned not only their house but also their life, their children, and society in general. Students may want to compare Hopper's house with the ones described in Oates's essay—does the painting try to evoke these same kinds of associations and emotions?

METHOD

Oates first mentions evolution early on in the essay when she dismisses artful treatments of abandoned houses in favor of "Darwin's great vision of a blind, purposeless, ceaseless evolutionary process." She also notes that the concepts of memory, womanhood, and self-identity evolve—they are always changing and adapting in order to ensure their survival. Later in the essay, Oates wonders what evolutionary advantage there could be in staying in an abusive, destructive environment such as the Weidel household; she figures that it may be an essential part of the feminine mystique. Ruth Weidel's cryptic comment "They all just went away" indicates that there was no real reason or purpose behind the decision to leave—it was simply a temporary strategy to escape extinction.

MEDIUM

Oates finds that the images created by Hopper and his fellow painters aren't true to the feeling of the house, that they've been idealized or romanticized in some way. She writes that "the real is arbitrary," that no artistic vision can capture the full effect because those visions are selective—they focus on some aspects above others. Oates's appreciation is for the chaos of reality above all else, "what assaults the eye before the eye begins its work of selection."

ADDITIONAL WRITING TOPICS

1. If for some reason your family home were to be abandoned, what sorts of items might a stranger find there? How might someone who never knew you or your family interpret these items?
2. Compare Edward Hopper's painting and Joyce Carol Oates's essay. Both works of art attempt to convey the emotions associated with the events they depict, but because they are presented in different media, the artists' approaches differ. What specific techniques do the artists use to present their respective audiences with the image of an empty house?
3. At the end of her essay, Oates says of her asking Ruth to come over for dinner: "my mother would have been horrified and would have forced me to rescind the invitation." Why do you suppose Oates feels her mother would have reacted this way? What details about the mother can you find throughout the essay to support this?

DIVERGENCES

Print

Levin, Gail, ed. *Silent Places: A Tribute to Edward Hopper*. New York: Universe, 2000. This is a collection of short fiction inspired by Hopper's paintings, etchings, and drawings. Students could consider how these writers translated Hopper's paintings into a new medium.

Oates, Joyce Carol. *Blonde*. New York: Echo, 2000.

———. *Broke Heart Blues: A Novel*. New York: Dutton, 1999.

———. *Them*. New York: Vanguard, 1969.

Web

http://www.artcyclopedia.com/artists/hopper_edward.html. Artcyclopedia's bibliography of online sites and articles includes this page on the art of Edward Hopper.

http://storm.usfca.edu/~southerr/jco.html. The University of San Francisco's Celestial Timepiece is easily the best collection of Oates materials online, with several pictures, detailed bibliographical information, awards, and of course selections from her various writings.

Audio/Visual

Blonde. Starring Patrick Dempsey. 4 hours, not rated. CBS, 2001. This recent made-for-TV movie is based on Oates's best-selling novel.

UNSPOKEN
David Sedaris

MEMOIR, 3 PHOTOGRAPHS, LETTER

The smiling apple-faced families in the stock photographs accompanying the David Sedaris selection do not appear to be the sort of family Sedaris grew up in. In "Ashes," Sedaris presents an image that is sharply different from such idealized visions. The Sedaris family knows that, in reality, families are neither perfect nor always happy, though they occasionally act the part in public or around friends. Sedaris and his family would rather steer clear of hypocrisy by developing a dry, dark sense of humor. Ironically, though, their choice to embrace sarcasm and irony as a way of relating to one another is just as much an act; faced with the tragedy of their own mother's death, the family members can't seem to express themselves outside of their cloud of dark humor, and by the end of the memoir David expresses some remorse for this.

One major topic you may want to cover in this selection is humor. Ask students to describe the Sedaris family's particular sense of humor. Ask them as well to identify situations where it would be appropriate or inappropriate. What other kinds of humor are they able to identify (e.g., slapstick, nonsense, mocking), and in what situations would it work? In this line of questioning, ask students to consider also the type of medium in which different situations could be presented—sight gags or visual humor, for example, likely would not translate well to the radio.

Another big theme tackled in this selection is the cliché, something for which Sedaris has little fondness. Try having students first offer their own definition of "cliché," and then compile a list of as many clichés as they can think of. After reviewing the list, you can agree to banish all of the items on it from use in any future writing assignments. You could also spend time helping students learn how to amend clichés so that they once again express meaning.

Direct students' attention to Sedaris's letter to his friend Sarah Vowell. Sedaris is in a gay relationship, and the way he characterizes his partner Hugh (crying at sappy romantic movies, throwing a wineglass at David during an argument) seems to satirize stereotypical straight relationships. Ask students to look for other allusions to heterosexual conventions in this letter.

CALLOUT QUESTION

This question invites a discussion of why we are susceptible to the clichés of romantic Hollywood films. Audiences flock to see films where the leading male returns for his true love at the last minute, where the heroine decides not to marry her obnoxious fiancé and instead runs off with the poor-but-noble hero, or where the faithful-but-doughty companion suddenly becomes "something more." Students could explore this cinematic mythology in terms of how it teaches us to understand our culture—you might direct them to notice what types of relationships are valued in these typical films, what counts as normal. Fairly recent films such as *Pretty Woman*, *When Harry Met Sally*, *Sleepless in Seattle*, *The Wedding Singer*, and *Pearl Harbor* prove that these conventions are still very much with us today.

MESSAGE

Sedaris learned his sarcastic manner from his mother, and his choice to use comedy is in keeping with his cynical disregard for all things cliché. It is part of his upbringing, his memoir suggests, to belittle sentimentalities that are corny and hokey; this point of view, as well as a quick wit and a sharp tongue (his humor is often crass), were gifts from his mother. He does, however, end "Ashes" on a somber note, a rare gesture on his part that gives the readers a sense of his true feelings—though he jokes about it, we know in the end that his mother's death did upset him.

METHOD

Sedaris tries to avoid conversations that are maudlin or sappy, because he feels these moments lack a certain sincerity or depth of genuine emotion. Instead of resorting to the stock phrases you might find on a greeting card, Sedaris and his family would rather stick to sarcasm. It would make a fun in-class exercise to have students research or generate from memory the kinds of overly sentimental phrases we associate with the greeting-card industry. Finally, in the letter to Vowell, Sedaris only indirectly quotes Hugh, a choice that makes him more pliable as a character—he can't speak his own words to defend himself against the fun that Sedaris pokes at him.

MEDIUM

Students should pay attention to Sedaris's use of dialogue throughout, which makes his piece multi-voiced and gives it the extra inflection of other people's style of humor. Also, there are shifts in tone, such as the somber reflection of the final two paragraphs of the piece, which suggest a different, more serious, manner of intonation. Sedaris also uses ellipses to note pauses.

ADDITIONAL WRITING TOPICS

1. Much of the humor David Sedaris and his family use to relate to one another is contained in running jokes that only they know about and find humorous. What sort of in-jokes do you share with your family or friends? Explain the humor behind them as well as how they came into being.

2. Characterize the relationship of the Sedaris family. Do they love one another? How can you tell? How does the family image David Sedaris creates compare to the stock photographs of happy families included in this selection?
3. Analyze Sedaris's writing style. How does he construct humor in his essay? What techniques, specific words or phrases, or scenarios does he use? Is the humor in his memoir about his family different from that in the letter to his friend?

DIVERGENCES

Print

Sedaris, David. *Barrel Fever: Stories and Essays*. New York: Little, Brown, 1995.

———. *Holidays on Ice*. New York: Little, Brown, 1998. Most of Sedaris's books are also available in audio format.

———. *Me Talk Pretty One Day*. New York: Little, Brown, 2000. Sedaris's latest book is a collection of humorous essays.

Web

http://www.bookpage.com/0006bp/david_sedaris.html. This page features a short BookPage interview with David Sedaris.

http://home.pacifier.com/~paddockt/sedaris.html. "The Unofficial David Sedaris Internet Resource" includes excerpts, biographical information, photos, fan letters, and tour information. Students unfamiliar with Sedaris can get an impression of his popularity here.

http://www.npr.com/. Students could explore the National Public Radio home page in order to get a sense of the various audience types to which Sedaris's material appeals.

http://www.salon.com/directory/topics/david_sedaris/. This list of past interviews and stories on Sedaris from *Salon* includes audio files of some of his readings.

Audio/Visual

The David Sedaris Box Set. Time Warner Audio Books, 2000. This is a large collection of some of Sedaris's more famous bits, from the early "SantaLand Diaries" to excerpts from his latest book, *Me Talk Pretty One Day*.

THE KITCHEN TABLE
Carrie Mae Weems

4 PHOTOGRAPHS

In these selections from her *Kitchen Table Series*, Carrie Mae Weems attempts to convey a narrative through the medium of photography; what this narrative is, though, is open to debate. This isn't meant to imply that photographs cannot tell stories, but the sequential ordering of the pictures in a series seems more in line with cinema than with narrative fiction. You might consider talking with your students about Weems's use of juxtaposition as a technique that complicates conventional ideas about photography. Rather than trying to capture a complete meaning in a single picture, Weems presents us with a meaning that doesn't take its full form until we experience the entire series. To illustrate this phenomenon, have students look at the pictures first in isolation and then as a series, and construct narratives for each. Note how the stories change when more information is given to the viewer.

Because Weems's pictures are so visually sparse, it is worth it to examine each picture closely. Try looking at the photographs in terms of their compositional value, their symbolic value, or their literal meaning within the narrative they help compose. For example, the kitchen table itself can be read as the compositional center of each picture; it can also be seen as a symbol for learning, as it holds an open book, pens, and paper; it also may be the site where a mother and daughter struggle over completing the girl's homework assignment. Students should focus on the changing positions of the figures as well. For instance, does the presence of the other adults at the table in *Untitled #2445* happen before or after the episode with the young girl, and how does the meaning of the story change in each case? When dealing with photos that are so similarly composed, the smallest of changes becomes monumental and heavily imbued with meaning.

MESSAGE

Students are likely to offer a variety of plausible narratives to accompany these photos, but you might point out to them the associations the kitchen historically carries as a domestic space. Further, the setting serves as a common denominator for the viewer, who remains familiar with the issues that unfold in the miniature dramas. Though the characters in the photographs do not appear to be communicating directly (they sometimes make eye contact, but no one appears to be in the middle of speaking), a definite tension is being communicated to the viewer.

METHOD

Each image offers the viewer the same perspective, and the same props often surround the table: chairs, light, book, paper. The continuity of the objects suggests a continuity among the photographs themselves—that they happened sequentially and tell a story. Compositionally speaking, elements such as empty chairs and the overhead light provide negative space that frames the center of the pictures, focusing the viewer's eye on the table itself and the action occurring there. The table protrudes into the foreground, thus placing the viewer at its head, making him or her a part of the picture, in a sense.

MEDIUM

The sparse decor of the kitchen suggests that Weems might be thinking of a particular socioeconomic class, and her depictions of African Americans suggests an ethnic or cultural specificity, but many students will identify with the kitchen-table woes regardless of class, race, or gender. In an era when the technology of photography has been surpassed with home video equipment and webcams, students might consider the relative accessibility of technology to particular cultural groups and, conversely, technology's universal appeal.

ADDITIONAL WRITING TOPICS

1. What noteworthy or memorable events have occurred around your kitchen table? Construct a detailed narrative in which you recount one or more such events, and how the setting affects your memory.
2. Spend a weekend taking candid snapshots of your choosing. Trade these pictures with a partner and construct a possible story based on their sequence.

3. In the headnote to this selection, art curator Dana Friis-Hansen is quoted as saying that Weems's pictures rely on "the modern viewer's media literacy, as we recognize the plot lines through our familiarity with a range of narrative forms from comic books and photo-novellas to soap operas and sitcoms." What does she mean by this? Provide an expanded definition of media literacy, using examples from the visual media suggested by Friis-Hansen to illustrate your points.

DIVERGENCES

Print

Patterson, Vivian. *Carrie Mae Weems: The Hampton Project*. New York: Aperture, 2001.

Sills, Leslie. *In Real Life: Six Women Photographers*. New York: Holiday House, 2000. This volume looks at Weems in relation to her contemporaries, including Cindy Sherman and Lola Alvarez Bravo.

Web

http://www.pbs.org/conjure/cm.html. PBS's online profile of Weems is part of the "Conjure Women" series.

http://www.universes-in-universe.de/america/us_afro/weems/weems1.htm. This online interview with Weems was conducted by Kaira M. Cabañas of Universes in Universe, an Internet project developed by German art historian Gerhard Haupt and Argentinean artist Pat Binder.

Audio/Visual

Behind the Scenes with Carrie Mae Weems. Directed by Ellen Hovde and Muffie Meyer. 30 minutes, not rated. First Run Features, 1992. This short documentary, hosted by magicians Penn and Teller, discusses Weems's approach to her art and the intended content in her photography. Features cameos by William Wegman, jazz drummer Max Roach, and theater costume designer Julie Taymor.

Illusions. Directed by Julie Dash. 34 minutes, not rated. Women Make Movies, 1983. Dash's short film, set in World War II, tells the story of a young African American woman who provides the singing voice for white movie stars. Deals with culturally specific themes similar to Weems's photos.

PUBLIC APOLOGY
Bill Clinton

NEWSPAPER, SPEECH

Very likely, this controversial episode covered in this section will still be in your students' recent memories, and they might have strong personal opinions about it. Therefore, discussing this selection in class may require careful moderation. If you prefer, you could invite students to explain the reasons why they hold the opinions they do, or instead you might ask students to suspend their opinions so that they can better study President Bill Clinton's national address for what it is—one man's effort to survive a rhetorically demanding situation.

In his address, Clinton must apologize to the American public for behavior that was not only unbecoming a president but also extremely embarrassing. Clinton's goal, therefore, is to deliver an apology that rings sincere on the one hand and maintains his authority and presidential stature on the other. Have students keep these two goals in mind as they analyze the various strategies he uses in the address. They should consider the rationale behind such rhetorical choices as the use of "our" and "we" throughout, the connotative meanings of words like "spectacle," and the shift in the address's focus from apology to vilification of the investigation. A useful exercise might involve bringing in other presidential addresses for comparison, such as Richard Nixon's resignation speech or Ronald Reagan's address regarding the Iran-Contra affair. Many critics have touted Clinton's aptitude as a speaker; students could debate the relative effectiveness of his address versus these other efforts.

Finally, students could consider the *Boston Globe* front page in the context of the media firestorm surrounding the Clinton/Lewinsky affair. The *Globe*'s treatment of the story, with its sizable headline and photo positioned "above the fold," seems in keeping with such exuberance. Ask them, based on the content of the *Globe*'s front page, if the paper has a particular political bias. You could also direct them to find other media events associated with the affair (such as the televised commentary of the impeachment proceedings or published portions of Clinton's deposition) in order to identify common elements of language, design, and rhetorical bias.

MESSAGE

Clinton uses the ideal of privacy, an ideal that Americans generally prize and certainly understand, as part of the reason behind his decision to mislead the public. The strategy is meant to foster empathy on the part of the audience. The idea of a line separating public and private life is obviously an issue for strong debate among students; they may want to discuss how that line is drawn differently depending on a particular person's social status or prominence in the media spotlight.

METHOD

The address begins with a short apology and then shifts to an attack on the government's investigation of the affair. Any misgivings Clinton had may have been about his *rhetorical* performance, that his speech "misfired" in front of the American public because he chose to turn to this attack rather than remain apologetic. Clinton's call to "move on" was certainly one shared by a large segment of the public at the time, but because the words came from Clinton, some felt that it was a little disingenuous—he wanted to leave the situation behind in order to escape scandal, not to repair the fabric of the nation.

MEDIUM

Students might notice Clinton's diplomatic diction. Instead of choosing words like "affair" or "sexual," he opts for "inappropriate relationship"; rather than "lie," he chooses "mislead"—these choices acknowledge wrongdoing without necessarily conjuring up damaging or embarrassing specifics. As noted above, he also deflects criticism by changing topics in his speech, a move meant to bring his audience into his own frame of mind. The second-to-last paragraph is a plea for unity, a classic rhetorical gesture intended to form a common bond between speaker and audience—if we can identify with him, then we are less likely to condemn him.

ADDITIONAL WRITING TOPICS

1. After reading the transcript of Clinton's address, look at the *Boston Globe* front page. What sort of message is the page's design meant to convey? Is this different from the tone intended in Clinton's speech? How so?
2. Which aspects of your life do you consider public, and which do you consider private? Should certain people, such as celebrities and politicians, be expected to give up their right to privacy in exchange for fame, money, and power? How would you react if the private details of your life became public matters of interest?
3. Educated as a lawyer, Clinton defended himself against allegations of perjury by saying that he did not lie but instead only muddied the waters of truth by resorting to semantic quibbling (he paid attention to the very precise meaning of words). Have you or someone you know ever used this strategy in order to get out of trouble? If so, recount the event in a detailed narrative.

DIVERGENCES

Print

Merkl, Peter H. *A Coup Attempt in Washington?: A European Mirror on the 1998–1999 Constitutional Crisis*. New York: Palgrave, 2001. This recent book offers a non-American perspective on the Clinton impeachment scandal.

Web

http://www.ardemgaz.com/prev/clinton/aapaula14sidea.html. This page contains an Associated Press excerpt from President Clinton's deposition in the Paula Jones case.

http://www.pbs.org/newshour/character/links/nixon_speech.html. PBS Online's transcript of Richard Nixon's 1974 resignation speech may be used in comparison with Clinton's address.

Audio/Visual

Ronald Reagan: The Great Speeches, Volume 1. Speechworks. 75 minutes. Audio CD. This CD features audio recordings of eleven of President Reagan's most well regarded speeches.

To Die For. Directed by Gus Van Sant. Starring Nicole Kidman. 105 minutes, rated R. 1995. This dark comedy deals with celebrity in a somewhat twisted way, as a young news anchor (Kidman) seeks fame by committing scandalous crimes of passion. Addresses the blurring of the public/private line.

WATCH ME
Webcams

4 COMIC STRIPS, 3 SCREEN SHOTS

Generally, people tend to look down on comics as a "low" form of art, but illustrators like Garry Trudeau work to challenge that stigma. You might begin discussion of this selection by asking students to list words or phrases they associate with comic strips and comic art in general. These associations might suggest that they view comics as silly jokes, incapable of conveying serious messages—material for juvenile reading tastes. Then, ask them if—and if so, how—a comic strip like *Doonesbury* complicates this viewpoint. Have them research other examples of comic art that functions similarly.

You could also ask students to describe the layout of Trudeau's strips, particularly how the visual elements complement the delivery of the verbal text. They might take notice of such elements as the unrealistic drawing style (and Trudeau's trademark facial features and his lack of shading); the similar pacing of the jokes (the inciting dialogue presents the topic, which leads to some sort of problem between the characters, which is resolved in the punchline of the final panel); and the use of a silhouetted panel toward the end of the weekday and Sunday strips. Another class exercise could involve editing Trudeau's Sunday strip (because of different space considerations from newspaper to newspaper, syndicated cartoonists often draw their Sunday strips so that panels can be cut). If the strip were to conform to the four-panel weekday format, ask them which panels they would use and what criteria would guide their decisions.

Thematically, this selection is about the webcam phenomenon, a topic that Doonesbury treats humorously. The strips suggest that audiences are attracted to the random feel of webcam sites and that, once published online, even the most mundane of occurrences becomes a source of voyeuristic excitement. As students explore the material from JenniCam.com, ask them if they share in this fascination, and if they can explain it (some students may already be familiar with this site or similar ones). Have them take note of the site's design—the dark background, the closely cropped images on the intro screen, the placement of the small gallery thumbnail pictures and the comparatively larger cam shot. What different types of activities do we find Jennifer Ringley participating in while she is in front of her cameras? More to the point, why do we not consider her site pornographic even though sexual activity does occur? Does Ringley's writing correspond with the images of her archived on the site? Also, what do the banner advertisements say about the site's host and its intended audience?

MESSAGE

Trudeau's characters interact not like father and daughter, but almost as if the roles are equal or even reversed; the father Mike is semi-stern and a bit cynical, while the daughter Alex has a glib, distracted tone. In front of the camera, Mike is stilted and nervous, while the savvy Alex obviously "gets it." And whereas Trudeau shows us Alex trying to manipulate her site by making her father angry at her, we don't get to look behind the scenes of JenniCam.com in the same way. We can only assume that Jennifer Ringley's self-portrayal is authentic and honest.

METHOD

The distinction made is between an event shown live and as it happens versus one that has been recorded or edited in some way. Obviously, the comic strip does not exist in continuous time, but time passes as we move from one panel to the next, a convention of the medium's visual language. The cartoon, unlike the continuously running camera, only presents us with selected information in order to compose a joke. The camera shows everything in a seemingly random, unordered stream.

MEDIUM

Webcam sites like JenniCam.com tend to run continuously, a condition that appeals to the audience—their curiosity makes them visit often, wondering if something big will happen soon. In the *Doonesbury* strips, the somewhat curmudgeonly father, Mike, seems to voice the sentiments of Trudeau (consider the father/cartoonist parallel), especially since Trudeau is known for poking fun at cultural phenomena, which Mike also seems to be doing.

ADDITIONAL WRITING TOPICS

1. Have you ever read (or do you continue to read) comic books? If so, in a short narrative or journal entry, describe the types of comics you read as well as any memories you associate with them.
2. In one of the *Doonesbury* strips, the father says, "Isn't real life a little random to sustain audience interest?" Alex replies, "It's story arcs that are predictable and boring. Random ROCKS!" Based on the popularity of sites like JenniCam.com, why do you think random rocks?
3. Spend a few minutes surfing JenniCam.com. How "real" is the person presented on this site? When does she act as if she's aware of the camera, and when does she forget it's there? What does she do off camera, when she leaves her house? Write three detailed scenes featuring the "real" Jenni as you imagine her outside her Web site.

DIVERGENCES

Print

McCloud, Scott. *Understanding Comics: The Invisible Art*. New York: HarperCollins, 1993. McCloud's illustrated book explains that comics are a legitimate art whose history of merging words and images reaches back centuries. Garry Trudeau once said of McCloud's book, "Most readers will find it difficult to look at comics in quite the same way again."

Trudeau, Garry. *Duke 2000*. Los Angeles: Andrews McMeel, 2000. A collection of strips chronicling one of Trudeau's most famous characters making a run for the presidency.

Web

http://www.doonesbury.com/. The Doonesbury Electronic Town Hall has a thirty-year archive and a lot of topical content.

http://www.jennicam.com. The JenniCam site is still going strong.

http://www.unitedmedia.com/comics/. United Media's Comics.com site features archives of several strips, including *Peanuts* and *Dilbert*.

Audio/Visual

The Truman Show. Directed by Peter Weir. Starring Jim Carrey. 103 minutes, rated PG. Paramount, 1998. Videocassette. Main character Truman is unknowingly the star of a 24/7 television program showcasing every moment of his life.

DOCUMENTARY PHOTOGRAPHY
Mary Ellen Mark

3 PHOTOGRAPHS

The headnote accompanying this selection points to Mary Ellen Mark's desire to raise public awareness through her photography. Referring to the people she has photographed, she says: "What I want to do more than anything is acknowledge their existence." Mark's photos are compassionate portraits of real people. They are meant to evoke feelings of shared humanity in the viewer—these people are like us, after all. You might begin discussing this selection by having students list stereotypes associated with the homeless. Then, compare that list to Mark's particular example of homelessness in the Damm family; how well (if at all) does this list describe the family depicted here? What are the discrepancies between the list and the photographs? To what do students attribute these stereotypes?

Try reading Mark's pictures in contrast to Jacob Riis's *Homeless Boys*, a prototype of the documentary photograph genre. You could direct students to note how each photographer has chosen to capture the subjects. Riis's boys are almost romantically idealized; they appear in a centralized, triangular position as if posing for a painting. Their eyes are closed, thus conveying a sense of passivity and helplessness. As viewers, we are not meant to identify with these picturesque street urchins, but rather to look at them with a vague twinge of Dickensian sentimentalism. By contrast, Mark's pictures do not quite work as portraits because their composition is slightly skewed. For instance, in the family shot, the mother is posed at an awkward slant, and the car frame (literally and figuratively) appears slightly off-axis. In the picture of Chrissy Damm and Adam Johnson as well, their diagonal position serves to unsettle the static frame of the photograph. Furthermore, Mark's subjects make eye contact with the camera, disrupting any chance at sentimentalizing or objectifying the figures. They look directly at the camera, and so they look directly at us, which makes them active agents in the exchange between picture and viewer.

A final activity might involve discussing our fascination with people who exist on the edges of mainstream society. Mark's photography offers a relatively tame look at unconventional lifestyles, but why do popular talk shows, films, and even some of our greatest works of literature offer us glimpses of people who are down and out? In times of peace and prosperity, a society's art and culture often look inward for inspiration. Why does our society want to look at the representations?

MESSAGE

In *The Damm Family in Their Car*, the slouching bodies, piercing stares, and intense facial expressions indicate that this is a shot of a distressed family. The photograph's center is the intersection of the diagonal lines formed by Dean's arms and Linda's body, which reinforces their identity as both parents and husband and wife. The picture's composition is not unlike a family portrait, with children gathered around centrally positioned parents, but the surprising element is that the photo is framed by a car rather than a formally decorated room or a scenic backdrop familiar to portrait photography.

METHOD

At best, gender roles are unsteady because the Damm family is located outside mainstream society—the rules of appropriate husband-and-wife behavior don't apply. Dean and Linda are awkwardly posed in the family photo; they appear uncomfortable, and Dean's posture seems to suggest more of a protective attitude than warm affection. In contrast, the photograph of Chrissy and Adam appears more sexualized, with two lone reclining figures making aggressive eye contact with the camera.

MEDIUM

In the family shot, Mark has chosen to omit much of the North Hollywood scenery, perhaps because it would be too distracting in a compositional sense. Also, to include more of the background would diminish the focus on the Damm family, and the thematic impact of showing a homeless family in isolation from society would be lost. The close cropping of the shot with Chrissy and Adam excludes the squalor of the abandoned house in which they are squatting.

ADDITIONAL WRITING TOPICS

1. Compare and contrast Mark's pictures of the Damm family and Riis's picture of homeless boys. With regard to subject matter, composition, and tone, how are they alike? How are they different? How can you tell that Riis's picture is from the 1890s, while Mark's pictures are from a century later?
2. Some critics have claimed that Mary Ellen Mark's career involved the exploitation of homeless people for her own gain, that what she calls documentary photography is actually more like coerced or staged photography based on her relationship with the Damms. Do you agree or disagree with this criticism? Defend your answer.
3. Describe an encounter you have had with a homeless person. Include your behavior, any dialogue you exchanged, and your reaction afterward.

DIVERGENCES

Print

Mark, Mary Ellen. *American Odyssey*. New York: Aperture, 1999. This volume represents Mark's take on images of Americana.

———. *Streetwise*. New York: Aperture, 1988. Mark's collection of photographs of Seattle's teen drug scene has been highly acclaimed.

Riis, Jacob. *How the Other Half Lives: Studies among the Tenements of New York*. New York: Dover, 1971.

Web

http://www.maryellenmark.com/. The artist's home page includes an online gallery, biographical information, and a bibliography.

http://www.salon.com/people/bc/2000/03/28/mark/. The online magazine *Salon* presents an excellent profile of Mark's career.

Children with a Secret
Kathleen Coulborn Faller

ESSAY

It goes without saying that Kathleen Coulborn Faller's essay "Children with a Secret" deals with some potentially disturbing subject matter, so the utmost sensitivity should be used when introducing this selection to students. Focus writing and discussion on the piece itself rather than delving into peripheral subject matter. Begin by asking students to characterize Faller's tone in the essay and to explain how that tone relates to her position on the topic. For example, references to the "Backlash" (capital *B*), along with words like "obstacle" and "jeopardy," suggest a sense of urgency that Faller wishes to pass on to the reader.

The essay begins with a brief history of the cyclical shifts in cultural attitudes toward child sexual abuse; we tend to fall into periods when we accept that abuse exists and then into periods when we doubt its validity. Citing several recent studies, Faller argues that we are currently approaching the doubting stage once again. Her main argument is to convince readers that we should take accounts of abuse more seriously. To establish her authority on the subject, she recounts her credentials as a long-time evaluator of childhood sexual abuse.

The remainder of this essay is an analysis of the different strategies sex offenders use to intimidate victims, such as threats of physical pain or loss of family members and pets. This is followed by a series of graphic case studies meant to present the reader with the harsh reality of abuse. Faller concludes her essay by reiterating the purpose of raising social awareness, especially given the current social climate of doubt.

The dominant theme of this essay is, of course, telling secrets. You may want to ask your students who is telling the secrets and whose secrets they are: the children's? the abusers?

Structurally, this essay is a particularly good example of argumentative writing because its purpose, use of evidence, and writer's authority are clearly articulated and consequently easy to locate. Students can benefit from examining it closely, perhaps even constructing an outline of it for use as a model or template for their own persuasive writing.

MESSAGE

Students should be able to identify Faller's purpose in this essay as one of raising awareness about childhood sexual abuse, particularly with regard to understanding the child's point of view. The problem she faces in accomplishing this goal, as she suggests in her conclusion, is the growing cultural "Backlash" that threatens to enshroud the issue in a cloud of skepticism. The case studies she includes are part of her attempt to fight this skepticism.

METHOD

Faller's introduction makes her essay argumentative rather than simply expository. Here, she poses the problem (that the historical enshrouding of child sex abuse threatens to return) and indicates how her essay will attempt to address it (by educating readers as to the various dynamics involved in sex abuse case studies).

MEDIUM

Students might suggest a variety of reasons that children keep sexual abuse a secret, including but certainly not limited to confusion, an incomplete understanding of social taboos, or loyalty toward the abuser.

ADDITIONAL WRITING TOPICS

1. Based on Faller's case studies, describe the strategies she employed in order to get abused children to tell their secrets. In your opinion, why were these techniques effective? What other strategies might also work?
2. At the end of her essay, Faller mentions a "Backlash" in our society's attitude toward the idea of sexual abuse, an attitude marked by an increasing level of skepticism. Do you agree with her assessment? Defend or refute this position, providing appropriate evidence to support your position.
3. Analyze "Children with a Secret" in terms of its rhetorical structure. What counts as evidence for Faller? What kind of diction does she use in constructing her argument? Where does she rely on logic to persuade her audience, and where does she appeal to the reader's emotions?

DIVERGENCES

Print

Faller, Kathleen C. *Child Sexual Abuse: An Interdisciplinary Manual for Diagnosis, Case Management, and Treatment*. New York: Columbia University Press, 1988.

———. *Child Sexual Abuse: Intervention and Treatment Issues*. Washington, D.C.: U.S. Department of Health and Human Services, 1993.

Web

http://www.civitas.org/. CIVITAS is a nonprofit organization that produces communication materials on issues related to caring for children.

Audio/Visual

Interviewing for Child Sexual Abuse: A Forensic Guide. Produced and Directed by Kevin Dawkins. 35 minutes, not rated. Guilford Publications, 1998. This video features Kathleen Coulborn Faller giving advice on effective interviewing techniques for those working with children.

The Problem of Our Laws
Franz Kafka

SHORT STORY

Much of Kafka's writing concerns individuals facing an absurd world full of nameless power and fear, what some might call a "dystopia," the opposite of a utopia. You might begin exploring "The Problem of Our Laws" by asking students to list other examples of dystopias they have encountered in literature and film. Some common examples include George Orwell's *1984*; Anthony Burgess's *A Clockwork Orange* (and the Stanley Kubrick film adaptation); Ridley Scott's *Blade Runner*; Ray Bradbury's *Fahrenheit 451*; and, more recently, *The Matrix*. Through naming characteristics common to these types of texts, students should come to understand Kafka's short piece in terms of the genre—as a portrait of a bleak totalitarian landscape wherein the freedom of individuals is stifled to ensure the wealth and power of a ruling class.

Additionally, you should have students analyze Kafka's style in this piece. For instance, why is he intentionally vague, offering no mention of a specific place, date, or names? For what purpose does he use first-person plural pronouns such as "we" and "our"—to whom is he talking, and how is the reader meant to stand in relation to this audience? How might unusual word choices ("prodigious," "meager," "gainsaid," "repudiate") be explained?

As a final topic for consideration, you might note that this piece is a translation, a fact that could invite a lively discussion about the nature of art and what happens to it when it has been translated, adapted, or in some other way reproduced. Can we still find value in Kafka's writing after it has been taken out of its original German and away from its original context? Or does good art have a timeless, universal quality that makes Kafka's translation still work for us because the artistic *idea* remains intact?

MESSAGE

Kafka's piece is characterized by mostly simple language, and some might find its plain tone similar to that of political pamphlets, such as the ones written by Thomas Paine to advance the cause of the American Revolution. Students might debate the specific genre of the piece, because its lack of detail gives it no obvious defining characteristics. It might be considered along the lines of an allegory or a satire in the sense that it exaggerates an existing situation in order to ridicule it.

METHOD

That the narrator remains anonymous seems fitting given the subject matter—the discussion of a strict and oppressive government. The speaker's call to revolt is meant to be made public so that the people of this land can read it; because the nobles might read it as well, anonymity offers protection from harsh punishment. Also, the piece makes no real mention of time or place, a point that may be intended to suggest a certain universal attitude toward oppressive governments.

MEDIUM

Again, Kafka's vagueness indicates no particular regime, and maybe not even a particular type of government; thus, the piece can be read as a condemnation of "hidden" laws wherever they might be encountered—students today, for instance, may think of the political power of multinational corporations or the informal rule of law imposed by an urban street gang.

ADDITIONAL WRITING TOPICS

1. Are there laws in your town, state, or country that you disagree with? If so, choose one and explain why you find it unjust. How would you suggest overturning or amending this law?
2. Write a short response to Kafka's "The Problem of Our Laws" from the point of view of the nobility he criticizes. Argue for the necessity of keeping laws hidden from the public. How does this help society function better?
3. This short story is originally from the collection *The Great Wall of China: Stories and Reflections*, whose title alludes to a specific type of empire. What national governments could Kafka's piece apply to today? How are these governments characterized in our society?

DIVERGENCES

Print

Kafka, Franz. *Amerika*. Translated by Willa and Edmund Muir. New York: Schocken Books, 1996. The Muir translations are generally regarded as the best English versions of Kafka's work.

———. *The Castle*. Translated by Mark Harman. New York: Schocken Books, 1998. This is a new translation based on the restored text.

———. *The Metamorphosis, In the Penal Colony, and Other Stories: With Two New Stories*. Translated by Joachim Neugroschel. New York: Scribner, 2000. This recent collection contains previously unpublished work.

———. *The Trial*. Translated by Willa and Edmund Muir. New York: Knopf, 1972.

Orwell, George. *1984*. London: Secker & Warburg, 1987.

Web

http://www.cs.technion.ac.il/~eckel/Kafka/kafka.html. This site has several good-quality pictures of Kafka.

http://www.pitt.edu/~kafka/intro.html. This University of Pittsburgh site is dedicated to Kafka studies.

Audio/Visual

Blade Runner. Directed by Ridley Scott. Starring Harrison Ford. 117 minutes, rated R. Warner Home Video, 1991 (filmed in 1982). Videocassette.

A Clockwork Orange. Directed by Stanley Kubrick. 137 minutes, rated R. Warner Home Video, 1991 (filmed in 1971). Videocassette.

The Matrix. Directed by the Wachowski Brothers. Starring Keanu Reeves and Laurence Fishburne. 136 minutes, rated R. Warner Home Video, 1999. Videocassette.

STATE SECRETS
My Lai

TRANSCRIPT, JOURNAL, 3 PHOTOGRAPHS

It is likely that today's students will not be familiar with My Lai or with other aspects of the Vietnam War. Therefore, it might be a good idea to offer a context for the images, transcript, and journal excerpt included in this selection. Have them look for other Vietnam-era texts online or in the library so that they can get a sense of the divisive public sentiment surrounding the war, as well as the news media's participation in that divisiveness. Seen in this light, Ronald L. Haeberle's documentary photographs are not the records of a job well done, nor are they simply intended as evidence to expose the aftermath of one terrible event; they are, in fact, commentaries against the very war itself.

If you wish to stay within the confines of the book, the pieces in this selection are particularly good for comparing the way each medium attempts to portray a specific event as "true." Without launching into a heady philosophical discussion of what truth is, you might ask your students how each piece gives us a different side of the truth. Thomas R. Partsch's journal, candidly written, was meant to be kept private, and therefore Partsch did not censor himself and the events appear to be more or less described as they happened. William L. Calley's testimony was delivered under oath by a man trained to respect the sanctity of his country's institutions; we are therefore likely to trust his words and even to sympathize with him when he says that the violent loss of his troops led to his destructive emotional state. Haeberle's photographs stand as irrefutably graphic recordings of the event as it happened; it is nearly impossible to ignore the terrified expressions on the victims' faces or the disturbing, unnatural poses of the slain children.

Finally, you could have your students tackle the topic of how atrocity is represented in the news media today, and if these representations are different from those of three decades ago. Have them research tragic events covered in the news in recent years, such as the Columbine High School shootings or the Oklahoma City bombing, and debate whether the media want simply to inform viewers and raise their social awareness or to get higher ratings.

MESSAGE

The hurried and candidly honest tone make Partsch's journal entry appear authentic. The spelling and grammatical errors serve to reinforce that he wrote this to keep an account of events for his own records and not necessarily for any artistic reasons. The matter-of-fact attitude displayed in the entry is that of a soldier who, although he admits that he did not participate in the massacre, did not exactly condemn the actions either.

METHOD

As a class exercise to explore the transcript of Calley's court martial, one group of students could construct a prosecuting case and the other group could make the defense. In composing their strategies, students would have to question the reliability and authority carried by each type of evidence. They would also have to consider ways in which that evidence could be refuted or rendered invalid (i.e., though Haeberle's photographs stand as a more or less "true" documentation of the event, it could be suggested that they were staged or somehow doctored).

MEDIUM

In different ways, each document reinforces different aspects of the massacre at My Lai. Haeberle's pictures seem the most disturbing because we get glimpses of the victims themselves—the tortured faces, the sheer number of crumpled bodies. Partsch's journal entry is shocking, too, because we get insight into the mental state of one of the participants—the apparent lack of horror in his words is unsettling. Since My Lai, combat photography has created a kind of checks-and-balances effect: soldiers are aware that their actions may be recorded for public consumption. In another sense, the sanitized war aesthetic of World War II returned in a different form in the Persian Gulf War, during which the media showed no real scenes of combat but rather countless video feeds of smart bombs successfully reaching their intended targets.

ADDITIONAL WRITING TOPICS

1. Search out media depictions of the Persian Gulf War and compare them with images from the Vietnam War (such as Ronald Haeberle's photographs), pointing out notable similarities and differences. Show how the different techniques reflect our nation's attitudes toward each of these wars. (While several polls show that the majority of the American public supported the Persian Gulf War, the Vietnam War was steeped in conflict.)
2. There has often been heated debate about the ethical obligation of photojournalists. Photojournalists talk about the duty of capturing events as they happen, which means they should not interfere in the action. Some critics say that this hands-off policy is cruel to the victims of war, especially when the photographer has the resources to help someone in trouble. Which side of this issue do you support, and why?
3. Based on William Calley's court-martial testimony and Thomas Partsch's journal entry, can you explain the motivation of the soldiers involved in the My Lai massacre? What was it about their feelings toward the enemy, their lost friends, and war in general that led to this event?

DIVERGENCES

Print

Baudrillard, Jean. *The Gulf War Did Not Take Place*. Translated by Paul Patton. Bloomington: Indiana University Press, 1995. Cultural scholar Jean Baudrillard's provocative piece suggests that the majority of the Persian Gulf War was staged by the news media.

Web

http://www.loc.gov/exhibits/treasures/trm003p.html. This Library of Congress site details the development of the Vietnam Veterans Memorial.

http://thewall-usa.com. Among its numerous features, the official site for the Vietnam Veterans Memorial contains a photo gallery, a slide show, and extensive links.

Audio/Visual

Casualties of War. Directed by Brian Depalma. Starring Sean Penn, Michael J. Fox. 2 hours, rated R. 1989. Videocassette. Feature film about a young American soldier's moral crisis when his platoon captures and abuses a Vietnamese woman.

The Killing Fields. Directed by Roland Joffé. Starring Sam Waterson, Haing S. Ngor. 142 minutes, rated R. Warner Studios, 1984. Videocassette. This is a film adaptation of Cambodian journalist Dith Pran's four-year ordeal as a prisoner in a Khmer Rouge concentration camp.

Maya Lin: A Strong, Clear Vision. Directed by Frieda Lee Mock. Not rated. Ocean Releasing, 1994. Videocassette. This documentary features the sculptor responsible for the Vietnam Veterans Memorial.

3 Shaping Spaces

Before your students read this chapter, you may want to discuss with them the concept of "space." While some students may be aware of space used in the conceptual sense, some may find difficulty with the more abstract connotations of the word.

Discuss with students how Mr. Lee, a homeless man, feels the human desire to create his own space and so creates a makeshift house, a unique, endearing home he can call his own. It is almost as if, without this space, Mr. Lee is unable to see his place in the world, or in the cosmos—there is no relationship between the larger world, the macrocosm, and his own corner of it, the microcosm. Mr. Lee's creation of space enables him to understand himself; with it, come his large signs and elaborate, artistic knots, distinctly signaling that this is his home.

Help students relate to this by asking them to describe their dorm rooms or rooms at home—the specific decorations, furniture, and personal belongings in the room. Then ask: How does the room and its furnishings represent you, the owner, and what about you do they fail to express? Also encourage students to think of spaces they have visited—buildings in their hometown, a park in a large city, a beach on a faraway island—and ask them what these spaces have meant to them.

A Nice Place to Stay on the Internet
Yahoo!

BILLBOARD

In the relatively short existence of the World Wide Web, Yahoo! stands as one of the elder statesmen of the medium. Having begun in 1994 as a way for two Stanford graduate students (David Filo and Jerry Yang) to catalog their favorite sites, it quickly became one of the most popular spaces on the Web as a starting place for many first-time users. Yahoo! does have its detractors, though; some surfers displeased with the site have unofficially said that the name is an acronym for "Yet Another Hierarchical Officious Oracle," and on occasion hackers have defaced the site's home page. You might discuss with your students the clique-like quality of online communities and, in particular, the divide between casual users who visit sites such as Yahoo! and computer-savvy users who look down on "newbies." Try visiting various discussion sites, both pro- and anti-Yahoo!, to help characterize this divide.

Beginning with the comment by technology pundit Jonathan Koppell, discuss the concept of the World Wide Web as a space. Koppell points out that the grammar and vocabulary associated with cyberspace reinforce the feeling that we are moving through space as we interact with the medium. Koppell offers a list of spatial language that students can build on when discussing this topic. Also, ask them how this spatial component is reinforced by graphical elements in cyberspace, such as the icons of most browser applications (Netscape's ship imagery or Internet Explorer's earth-based logo, for example). As a point of contrast, note that search engines like Yahoo! are subject to the limitations associated with thinking about cyberspace in terms of physical space. By design, they work by collecting disparate links on a topic that a user might otherwise not be able to find in proximity to one another.

Finally, tackle the question of why we define this medium as a space—do we have a need to locate ourselves within this virtual environment so that we can feel like we control it? What does this say, if anything, about the natural environment?

MESSAGE

The billboard's design hearkens back to the aesthetic associated with 1950s Las Vegas—flashy and hip. It mimics a hotel sign and promises a vacancy for the potential visitor. Moreover, this virtual vacant room is pitched to the viewer as a "nice place," a pleasant, familiar, and friendly alternative to the impersonal chaos associated with the rest of cyberspace.

METHOD

The location of this billboard reinforces the kitschy, retro feel of the design. The space-age 1950s aesthetic finds itself at home above the type of store we associate with that car-obsessed time, and the smallish auto store is dwarfed by the billboard, emphasizing Yahoo!'s message. Furthermore, the ad echoes the fascination our culture felt for the future during the middle of the twentieth century (think of the television cartoon *The Jetsons*, for instance), a fascination matched in intensity by our current interest in the online world—a new kind of space.

MEDIUM

Yahoo!'s use of the billboard (a traditional advertising medium) might be an acknowledgment of that medium's strength in reaching large numbers of potential customers. This is something that online advertising can't exactly duplicate, because it involves active participation on the part of a user to visit a site in order to see a banner ad. By contrast, a billboard doesn't rely on active viewers—anyone can catch a quick glimpse of it as he or she drives to work every morning. Like so many advertising strategies, this one relies on ambiguity to target potential customers—the 1950s aesthetic may attract those who are directly familiar with it as well as younger people who may understand it as ironic or retro.

ADDITIONAL WRITING TOPICS

1. Spend some time studying Yahoo!'s branding strategy. Consider such features as the choice of font and color, use of punctuation, and where the logo is placed throughout the site. Explain how you see the logo functioning—how does it represent the company's image, what audience is it trying to reach, and what slogans are used? If you were in charge of redesigning Yahoo!'s site, what other names and logo designs would you use instead?
2. Recall your first experience with the Internet. Describe the feelings you experienced as you interacted with this new medium. What sites do you remember visiting, and how were they designed? How has your experience of the World Wide Web changed since this initial contact?
2. Compare and contrast Yahoo! with other portal-based search engines such as Excite.com or About.com. In terms of overall design, image, and the various features each site offers, how are they similar? How are they different?

DIVERGENCES

Print

Anderton, Frances. *Las Vegas: The Success of Excess*. London: Ellipsis London, 1997. This study of the architectural aesthetic of classic Las Vegas gives attention to the use of signs and billboards.

Web

http://docs.yahoo.com/info/misc/history.html. Part of Yahoo!'s media relations package, this page offers a history of the site since its inception in 1994.

http://www.excite.com; http://www.about.com. These are two of the search engine portals similar to Yahoo!

Audio/Visual

Tron. Directed by Steven Lisberger. Starring Jeff Bridges. 96 minutes, rated PG. Disney, 1982. Videocassette. This cult-status Disney film offers an early imagining of what cyberspace might look like—full of neon lights, motorcycles, flying discs, excitement, danger, and love.

PHONE BOOTHS
A Place to Talk

PAINTING, ESSAY, PHOTOGRAPH

As the "Key Terms" definition in this selection suggests, Ian Frazier's essay is a good model of comparison/contrast organization. This is a good point from which to begin exploring the piece since, very likely, students will be familiar with this organizational method and have probably used it in their own writing. Also discuss Frazier's use of the anecdote, as well as the rhetorical value in drawing on one's own remembered experiences. Frazier values pay phones over cell phones because they are included in his own memories, and by extension he imagines others having had similar experiences. This argument, unlike traditional comparison/contrast writing, is not based on a reasoned assessment of the advantages and disadvantages of each technology but rather on the emotional associations Frazier himself makes with each type of phone.

Additionally, the essay deals with memory in a specific way. As a meditation, it starts from a remote and personal memory of calling his girlfriend in Florida from Montana, which leads to a flood of generalized observations about pay phones that Frazier has collected over the years. After several pages of waxing nostalgic about the pay phone and lamenting the popularity of the cell phone, he ends up sounding a larger note—the wish that physical spaces could mark the passions of humanity. Structurally, you could discuss the way this connecting of memories by association mirrors other strategies of organization from various media, such as hypertext links, cinematic editing, literary allusion, metaphor, or analogy.

When discussing Richard Estes's *Telephone Booths*, you can direct students in any of several ways. With regard to the painting's composition, note that the people in the frame are turned away from the viewer, and that the row of phone booths creates a balanced and uniform geometrical pattern. Also, discuss the role that reflection plays in the painting. Noting features like this can naturally lead to talk about the symbolic function of the painting—what is Estes saying about the urban world and the status of the people in that space? One question to ask about Estes's work is why he chooses to mimic photography—what is he saying about the media of painting and photography and their relation to one another? Is painting able to accomplish something artistically that photography cannot? Does painting need to borrow the aesthetic of photography in order to survive? To properly answer questions like this, your students may need to conduct some background research on the history of photo-realism. Finally, have students discuss Lauren Greenfield's photograph *Phone Booth* as a counterpoint to Estes's work, further supplementing this line of exploration.

CALLOUT QUESTION 1

Direct students' attention to Estes's realistic use of lines, shadows, and particularly reflections (like many photo-realists, he often paints reflective surfaces like chrome and mirrors). The orderly scene of booth after booth is almost too pristine, and it certainly offers an idealized vision of an urban scene. Consider that the medium of painting invites the artistic manipulation of images more readily than photography does.

CALLOUT QUESTION 2

Frazier's anecdotes help ground the object at hand in his personal experience, so that we know his essay is more than a simple report on the history of pay phones. By showing the reader how this ordinary, everyday device helped shape his memories and identity, Frazier communicates a feeling of nostalgia for the pay phone and the culture surrounding it.

MESSAGE

More than anything, Frazier wants to have physical monuments to his memories, and for him, the more ordinary these monuments are, the more authentic they are—this is why the pay phone stands as a prime example. As he puts it in the last paragraph of his essay, "Ideally, the world would be covered with plaques and markers listing the notable events that occurred at each particular spot." Frazier's title, a play on the phrase "dearly departed," is suggestive of the remorse he feels for the pay phone's demise.

METHOD

Estes's *Telephone Booths* showcases the anonymity of the urban space, as well as the isolation resulting from urban crowding. Multiple booths reinforce this idea, whereas a single booth with an impatient caller waiting outside it would give the figures in the painting a sense of individuality absent from the work as it stands. Frazier's main point in "Dearly Disconnected" is that he misses the role the pay phone played in his personal life; by making his essay personal, he shows readers the intimate link between technology and culture. In the more traditional comparison/contrast genre, this reliance on personal anecdote would be lost.

MEDIUM

Whether or not Estes's painting makes a valid social commentary could certainly be a topic for debate among a class. Students may want to consider the photo-realist movement itself, and whether or not paintings made to mimic the look of photographs are themselves some sort of comment on technology or the status of high art. By contrast, simply depicting things as they are seen may strike some students as both an empty social critique and a meaningless artistic statement. Some may feel that Frazier's essay is better at specifically articulating its message, while some may see value in Estes's ambiguous treatment of urban spaces, technology, and the lack of individual identity. To be sure, this is a complicated question, and one that invites students to consider other visual texts from the book (by Cindy Sherman, Mary Ellen Mark, or Weegee, for example) that also attempt to comment on society, but in similarly ambiguous ways.

ADDITIONAL WRITING TOPICS

1. In a short essay, choose a common appliance or device similar to the pay phone and use it as a symbol for recounting memorable events in your life. Think of items like personal computers, ironing boards, microwave ovens, or electric razors.
2. Frazier writes that the pay phone's "ordinariness and even boringness only make me like it more; ordinary places where extraordinary events have occurred are my favorite kind." Describe one thing you like that other people might find boring, or an otherwise ordinary place that you associate with an extraordinary event. Provide as many details as possible in your narrative.

3. Working in pairs, write a short dialogue that might take place between two people over the telephone. Compose the dialogue so that one character's lines are mundane and commonplace while the other character's lines are exciting and provocative. Be sure that the dialogue fits together logically, but be as creative as you'd like in constructing your scenarios.

DIVERGENCES

Print

Frazier, Ian. *Coyote vs. Acme*. New York: Farrar, Straus, and Giroux, 1996.

———. *Dating Your Mom*. New York: Farrar, Straus, and Giroux, 1986.

———. *Family*. New York: Farrar, Straus, and Giroux, 1994.

———. *Great Plains*. New York: Farrar, Straus, and Giroux, 1989.

———. *On the Rez*. New York: Farrar, Straus, and Giroux, 2000.

Meisel, Louis K. *Photo-Realism*. New York: Abradale Press, 1989. This is a complete history of the photo-realist painting movement with profiles of several of the main figures, including Richard Estes.

Web

http://artcyclopedia.com/artists/estes_richard.html. Artcyclopedia's entry on Estes includes a thorough catalog of online image galleries, interviews, and articles.

http://www.motherjones.com. Ian Frazier's essay was taken from *Mother Jones*, a liberal magazine featuring "news and resources for the skeptical citizen."

http://www.2600.com. The online hacker magazine *2600* features a section on "phone phreaking," which was a pastime of early hackers involving pay-phone piracy.

Audio/Visual

"Disappearing Pay Phones." National Public Radio, February 14, 2001. Recording. NPR's Adam Hochberg reports that BellSouth plans to abandon its stake in the pay-phone business by the end of 2002, which has some critics concerned. For tape or transcript, call 1-877-NPR TEXT (1-877-677-8398). Also available in RealAudio format on NPR's Web site, http://www. npr.org.

AMERICAN LEISURE
Mitch Epstein

2 PHOTOGRAPHS

Just as it did with to Richard Estes's painting of telephone booths in the preceding selection, the question of whether or not Mitch Epstein's photographs are intended as social commentary surfaces here. It can be a particularly challenging idea to suggest that photography has political potential—we tend to read pictures as blank or neutral records of real life, and many people believe that pictures capture events as they happened and nothing more. However, the case can be made that Epstein's pictures are political, and you may find it worthwhile to explore how they are staged or composed in order to emphasize the artist's point of view. In *Cocoa Beach, Florida*, for instance, the off-center angles of the vehicles, the figures curled up on the car hood, and the busy foreground work together to suggest a certain paradox between a growing culture of leisure and recreation in the 1970s and '80s and a reality of teeming, crowded RV parks and swimming beaches. In the caption to *56th Street and Fifth Avenue*, the statistics (originally included as part of the photograph by the editors of *Architecture* magazine) indicate a particular interpretation of the photograph as a comment on our hectic pace of life.

This selection invites a generalized discussion about what the purpose of art is in a society. Should art simply be something pretty to look at hanging on museum walls, appreciated solely because of the technical skill and talent of the artist? Or should it challenge the comfortable or commonplace ideas a society has about the way life ought to be? Can art ever be just one thing or the other?

MESSAGE

In *56th Street and Fifth Avenue*, the distanced viewpoint of the office building resembles a grid, and the people inside the offices retain a sense of anonymity. This is in keeping with the anonymous feeling associated with statistics, a form of evidence that doesn't take into account the opinions of individualized voices. Students may be compelled to read the photograph as an illustration of the statistics, trying to match up individual figures with the particular strategy they use to create more time for themselves.

METHOD

Epstein's recreational photograph is highly ironic, and the source of this irony is the juxtaposition of the wide-open sky in the top half of the frame and the extremely crowded bottom half. This contrast is intended as a comment on the packed, uncomfortable conditions of the recreational culture that reached critical mass during the mid-70s to mid-80s. This unsettled feeling is echoed in the props located in the frame—items such as cars, trailers, and lounge chairs are not only cropped but also positioned along vertical lines, which gives the composition a sense of movement that does not fit with traditional notions of leisure and recreation. The fact that the lounge chair is empty, and that the two figures are sleeping on the nearby car hood instead, drives the point home—none of these people are lounging.

MEDIUM

56th Street and Fifth Avenue offers us a perspective that we're not usually likely to see, and the result resembles not so much a cross-section of an office building as a series of panels assembled together in a montage fashion. This and other abstract tendencies emphasize the photograph's idea of person-as-statistic. According to the photograph and the statistics, individuals are not differentiated but rather anonymous background elements, secondary to the geometrical balance of the shot.

ADDITIONAL WRITING TOPICS

1. Define the terms "leisure" and "recreation," making sure to distinguish between the two. In your opinion, what activities and locations would be appropriate to associate with these concepts? Draw on your own experience when constructing your definitions.
2. Choose a single figure from either of Epstein's pictures and construct a fictional narrative about that person. Be sure to use other information from the picture to guide your writing.
3. Mitch Epstein was one of the pioneers of an art movement of the 1970s known as New Color

Photography, which sought to make color photography more artistic than it had been. Analyze Epstein's photography in terms of its artistic elements. For instance, you might consider his use of lines, contrast, shading, and (particularly) color. Do you find his photography artistic?

DIVERGENCES

Print

Epstein, Mitch. *The City*. New York: Powerhouse, 2001.

———. *Fire, Water, Wind*. Tenri-shi, Japan: Tenrikyo Doyusha, 1996.

———. *In Pursuit of India*. New York: Aperture, 1987.

———. *Vietnam: A Book of Change*. New York: W. W. Norton, 1996.

Web

http://www.outwestnewspaper.com/airstream.html. *Out West*'s site gives a history of one of Americana's enduring icons and a symbol for the recreational lifestyle—the Airstream Trailer.

Audio/Visual

Desert Blue. Directed by Morgan J. Freeman. Starring Brandon Sexton, Kate Hudson, Christina Ricci. 90 minutes, rated R. Columbia, Tristar, 1999. DVD. This bizarre tale of adolescent angst centers on a professor and his daughter's trip across America to shoot photographs of giant roadside sculptures.

AERIAL VIEW Perspective

ESSAY, 2 PHOTOGRAPHS, POEM

One of the prominent themes running through all of the pieces in this selection is that of change. For Pico Iyer, change happens to society at large, where the emerging global culture is incapable of holding the kind of local, nationalized perspective that people held before the birth of the "transit lounger." Alex S. MacLean's aerial photography depicts the sometimes-devastating changes in natural landscapes brought on by the ever-encroaching excess of civilization. John Updike's poem, with its two included revisions, represents the change of poetical thought itself—through these drafts, we see him struggle to find the perfect figure of speech to properly express his feelings for the Midwest landscapes viewed from above. Ask students to point out the different ways in which each artist depicts change in his work, as well as what the purpose of each work is.

The title of Iyer's essay, "Nowhere Man: Confessions of a Perpetual Foreigner," offers another good starting point for discussion. Why does Iyer think he's from nowhere, and what does it mean to him to be a perpetual foreigner? What is he confessing, and to whom? You could have students compare their own travel experiences with Iyer's—do they share his view that the new transit lounger is "unable to comprehend many of the rages and dogmas that animate (and unite) people"? Or do they instead see a value to traveling, a value that results in personal growth and more complete understanding of how other cultures live? Using this angle, stress the importance of being able to adopt various points of view as a valuable critical thinking tool.

When discussing MacLean's photos, direct students to the way he uses contrast in order to amplify civilization's intrusion into nature. In *McMansion and Woods*, for instance, the mansion is inserted in the middle of a patch of woods—the driveway makes a sharp incision into the land, and the boundaries of the estate are clearly defined. The other photo, *Automobile Junkyard Alongside River*, is literally divided in half, with the bottom half consisting of an unnerving number of junk cars and the top half of a calming, tree-lined river. Both of these photographs clearly maintain the division between what is natural and what is artificial, and MacLean himself suggests that we are meant to view the natural as the positive element in each image.

Updike's "Island Cities" offers a wonderful opportunity to impress upon students the importance of revision in their own writing. Not only do we see that even a writer of great renown has to rework his writing, but we are also able to glimpse the process itself. Students can examine the way Updike wrestled with his language, how he judiciously cut out phrases that didn't work for him, and how he came to revert to original phrasings. With earlier versions of the poem at their disposal, students can debate the validity of the choices Updike makes and even argue for changes of their own.

CALLOUT QUESTION 1

Just as there are some people who would condemn the photojournalist as passive bystander, taking pictures of war's atrocities without choosing to interfere in them (think back to Ronald Haeberle's pictures of the My Lai massacre in Chapter 2), Iyer's point is that the transit lounger is guilty of lacking humanity because of his or her passivity. The transit lounger is guilty of an indifference to nationalism, and, as the title implies, this essay is a confession for Iyer—he names his crimes in writing rather than offering up a safe, "objective" snapshot.

CALLOUT QUESTION 2

Iyer's final sentence emphasizes passive disinterest on the part of the transit lounger—an object going through a seemingly endless cycle of travel. The metaphorical connection between the transit lounger's life and the bag circling the carousel symbolizes the emptiness and rootlessness of Iyer's globe-trotting lifestyle.

CALLOUT QUESTION 3

As Updike says in the comment quoted in this selection, his purpose in revising was to be more selective in his choice of metaphors: "Adjudicating among metaphors, keeping some space between them, is one of a poet's responsibilities." Consequently, he prunes the language so that it is not overcrowded with concrete details (like the mention of St. Louis and Des Moines in the first line), and he chooses to use simpler language (the word "alkaline" in the third draft reverts back to "skyey blue" in the final version).

MESSAGE

Iyer's transit lounger is someone whose leisure time is spent mainly sitting while en route to his or her destination; this type typically frequents airports, planes, bus stations, and hotels near these depots. Though Iyer downplays the advantages of the lifestyle, he acknowledges the camaraderie among fellow lost souls; to avoid detracting from his main point, he does not suggest, for instance, that the transit lounger might genuinely benefit from learning about another culture.

METHOD

Much of Iyer's language reinforces the idea of not having a real sense of home; references to aliens, "a transcontinental tribe of wanderers," and nomadic travelers are among the most striking, as are metaphors of circularity and repetition, such as "pass[ing] through countries as through revolving doors" or the analogy of his life to unclaimed baggage circling the carousel in the last sentence. By contrast, Updike's metaphors are not so bleak. Comparisons such as "the main drag like a zipper," "dirt-colored cakes of plowed farmland," and quarries like "dewdrops of longing" have a certain energy and affirmation of life that Iyer's metaphors lack.

MEDIUM

Iyer is unquestionably critical of the aerial perspective because it erases one's ability to believe strongly in something. MacLean uses the aerial perspective in a negative way as well—to point out the harmful manner in which people exploit the environment through greed (the McMansion) or carelessness (the junked cars). Though MacLean and Updike both focus on documenting the impact of civilization from above, Updike offers a kinder perspective—to a degree, he romanticizes the strip malls, interstates, and quarries. An Updike-friendly photo from MacLean would probably not involve a distinct contrast between civilized and unadulterated landscapes, but instead would focus only on a landscape tamed by humanity.

ADDITIONAL WRITING TOPICS

1. Describe your hometown from above as it might appear to a passing transit lounger. What features of your town would stand out from an aerial view? What impression of your town would this view leave a stranger? What aspects of your town would the transit lounger not get a chance to experience?
2. Aerial photographer Alex S. MacLean refers to the large and expensive houses he photographs as "trophy houses" or "McMansions." Write a short interpretation of these names—what do they imply about the houses themselves, the people who own them, and MacLean's attitude toward them?
3. Analyze John Updike's revision process as it is shown in the facsimiles of his drafts of "Island Cities." Draft a poem of no more than one page and then revise it, keeping both copies. Provide written commentary with these versions explaining why you made changes, chose certain metaphors, or added text in your revision.

DIVERGENCES

Print

Iyer, Pico. *Falling Off the Map*. New York: Random House, 1993.

———. *The Global Soul: Jet Lag, Shopping Malls, and the Search for Home*. New York: Random House, 2000.

———. *The Lady and the Monk: Four Seasons in Kyoto*. New York: Random House, 1991.

Updike, John. *Rabbit Angstrom: A Tetralogy*. New York: Random House, 1995. This is a collection of Updike's four acclaimed "Rabbit" novels, started in the early 1960s and featuring main character Harry "Rabbit" Angstrom.

Web

http://www.lib.berkeley.edu/EART/aerial.html. The University of California Berkeley Aerial Photography and Satellite Imagery site includes digital facsimiles of several aerial views of the San Francisco area from throughout the 1900s.

http://www.uta.edu/english/V/test/agamben/v.1.html. Rhetoric scholar Victor Vitanza's Web presentation "Objects and Whatever-Beings: The Coming (Educational) Community" is a postmodern defense of students' adoption of the "whatever" position as a complex critical ability to accept multiple and conflicting viewpoints, much as Iyer's "transit lounger" does.

Audio/Visual

Fight Club. Directed by David Fincher. Starring Brad Pitt, Edward Norton, Meat Loaf. 139 minutes, rated R. 20th Century Fox, 1999. DVD. The story in this film revolves around Norton's character and his unstable sense of identity; his turn to a violent pastime is fueled in part by his constant airline travel and exposure to "single serving friends."

Remember When This Was Heavy Traffic? Negative Population Growth

ADVERTISEMENT

Students may be unfamiliar with the problem of global overpopulation, and also with the advocacy initiatives that attempt to deal with it. Have students inspect Negative Population Growth's Web site or that of a similar organization. Then, ask them to compare the ad in the text with the content they find on the site. It might make a good comparison/contrast exercise to look at NPG's content along with that of Physicians Against Land Mines (see Chapter 1); they could analyze each campaign's use of images and textual appeals to advance their respective causes.

In dealing solely with the ad, you could also direct students' attention to the different rhetorical strategies the ad employs in order to sway its readers. The first paragraph, with its allusions to a better America in the past, plays on readers' emotions, their sense of nostalgia; it assumes that increased traffic, crowded schools, and rampant development are conditions that the reader feels negatively about (and so there's no need to prove that point). By contrast, the second paragraph makes the logical case for supporting the organization's cause, citing several statistics from the U.S. Census Bureau. You might discuss these two distinct strategies in relation to the ad's overall purpose of soliciting support. For instance, are the statistics meant to persuade those readers who aren't convinced by the emotional appeal of the first paragraph, or are they simply used as part of a broader mission to educate the public on the group's cause? Also, which of the two strategies is more valued in this ad, considering their order, the headline, and the graphic layout of the page?

CALLOUT QUESTION

Rhetorically, the switch from "your" to "our" is a move that persuades the reader to identify with a group; in this case, the group is American people. While "your" implies a certain individuality (perhaps the position of the reader before considering the implications of overcrowding), "our" suggests a sense of community, moving the reader into a collective frame of mind that might make him or her more likely to take part in the problem's solution.

MESSAGE

NPG promotes an image of America that harks back to a past characterized by free and open spaces; by contrast, the image of the present and future is hinted at with negative language such as "overwhelmed," "outdated," and "unsustainable." The ad mentions the threat of overcrowded suburbs, the very locations to which middle-class Americans moved starting in the 1950s in order to escape dense areas in the first place. The use of the pronoun "our" in relation to "suburban communities" suggests that the ad is being targeted to these communities. The statistics pointing out the problem with immigration rates (and no direct mention of immigrants themselves) suggests that immigrants might feel excluded from this ad's intended audience.

METHOD

By not showing a literal depiction of overcrowding, the ad emphasizes the imminent, but still unseen, threat of overpopulation while focusing on an idyllic image supposedly related to the past. A road sign, like the statistics listed in the ad copy, is informational—it gives readers facts that dictate their immediate actions. Still, the sign's lack of uniformity (different fonts and awkward spacing, for no discernable reason) suggests that it is not a state-issued sign, but perhaps homemade, reinforcing the individualistic, community-minded message the ad conveys.

MEDIUM

The characteristically static images of NPG's print advertising (see Web site for other examples) might not translate well to a medium that uses moving images. Still, students could dissect the common features of NPG's ads, such as their minimalist connections between verbal and visual texts or their reliance on statistics, and suggest how the ads might look or sound translated to these different media.

ADDITIONAL WRITING TOPICS

1. In China, the government limits the number of children its citizens can have. From a logical standpoint, do you think this is a more effective way of controlling population than those strategies advocated by Negative Population Growth? Is it a more ethically responsible method? Explain your answers.
2. Imagine that you were responsible for NPG's ad design. If the intended effect is to scare the readers into making a donation to NPG, how would you rewrite the ad's copy? What if the intention is to make an appeal to the readers' sense of compassion? What photographs or other artwork might you use in each case?
3. Imagine what a world suffering rampant population growth might look like. Describe this condition in a brief but detailed scenario, considering the impact on individuals, governments, and the environment.

DIVERGENCES

Print

Hohm, Charles F., and Lori J. Jones, eds. *Population: Opposing Viewpoints*. San Diego: Greenhaven Press, 1995. This collection looks at both sides of the overpopulation controversy, with regard to appropriate solutions, the scale of associated problems, and the existence of a problem in the first place.

Web

http://www.npg.org. The Web site of Negative Population Growth contains a media kit, press releases, and data on overpopulation.

http://www.zpg.org. The Web site of companion advocacy group Zero Population Growth has content similar to that of NPG.

Audio/Visual

Human Growth. 20 minutes. Journal Films, 1994. Videocassette. This short video presents scientific discussions on the evolving rate of growth among the human species and on the effects of food supply on overpopulation.

LIFE IN MOTION
Nicole Lamy

ESSAY, PHOTO-JOURNAL

In "Life in Motion," Nicole Lamy starts out the road trip with her father with the assumption that she will be able to document a dozen past houses. She wished to share the photographs with her mother as a way to provide "straightforward narratives" of their lives. She seems to become increasingly aware of the futility of this goal, and as her pictures generate memories, literary allusions, and other bits of information, the reader comes to understand that Lamy's purpose in this essay is more than she originally proposes. Perhaps Lamy is trying to reach back to a pained childhood in order to repair relationships with both parents and bring a sense of closure to these issues in her own life. In the end, we are presented with the scene of a proud daughter handing a gift to her mother, who does not see a "logically ordered" catalog of a "largely unplanned" life, but rather evidence of several failures.

Have students study the text in order to describe Lamy's relationship with her father and mother. Note that she holds off mentioning her parents' divorce until well into the essay. Does her choice of language for each parent suggest that she feels differently toward them? Students should take note of Lamy's tone in describing her memories and try to infer whether or not she has fond recollections of her childhood. The scene of Lamy hidden away in her makeshift cardboard hut, her brief mention of growing up too fast, and repeated lists of lost objects all suggest that Lamy felt insecure about her highly mobile childhood. What else, then, could be the intended purpose of this carefully constructed collection of photographs that Lamy wants to share with her mother?

The photos accompanying "Life in Motion" offer students the opportunity to examine how media can document place. Have them compare Lamy's descriptions of the individual houses with their corresponding photographs. How do the descriptions reinforce the appeal of the photographs? How do the pictures themselves reinforce, or even challenge, these descriptions?

MESSAGE

Lamy wants to provide indisputable evidence of her memories; as she writes in the first section of her essay, she "wanted to gather the photos as charms against fallible memory." That fallibility is partially the result of her distorted view of things as a child, and partially of a more general distrust of the reliability of purely remembered events—she wants to document her past because it is more authoritative that way. She begins her essay with the belief that this is possible, but by the end she feels "ashamed" that she thought her project could objectively record the same set of memories for herself and her mother alike. This revelation comes about when her mother reads the pictures as documents of a series of failures rather than as evidence of a free-floating lifestyle.

METHOD

Lamy's claim that her photos "pretend no artistic merit" may be a false assertion, made to force the reader to return to the pictures with a more discriminating eye. The essay itself, with its penchant for lists and repetition of imagery, shows a conscious sense of thematic development. Consider the list Lamy provides in section 1 (purse, beret, silver dollars, negatives); the list of pets and pests that lived in their red-shuttered home (dog, rabbits, moths, and worms); and the list of rediscovered game pieces all as parts of an extended analogy for Lamy's lost life, which her photography journey is meant to rediscover. The allusions to classical sources—Cicero's mnemonic theory and the myth of Eurydice—seem to be in keeping with the epic quality Lamy wants to give to her journey from house to house.

MEDIUM

Lamy's approach to photography, as she describes it, is one of haphazard efficiency—she wants to take as many different shots as quickly as possible, and so she carelessly aims the camera from her hip or selects other nontraditional angles. Though this technique itself might be suggestive of a certain artistry, the real artistic statement takes shape as Lamy collects the pictures in an accordion book, coloring various parts of each house to show the human continuity among the buildings.

ADDITIONAL WRITING TOPICS

1. If you were to construct your autobiography around a set of photographs with a common thread, what kinds of pictures would you use? School portraits? Summer vacation shots? Explain your choice, and provide detailed descriptions of some of these photographs.
2. What is the significance of the essay's title, "Life in Motion"? Whose life is being referenced, and in what way is it in motion? Does the title suggest multiple meanings to the reader? Explain.
3. For what purposes could Lamy have set out on her photo-taking road trip, other than making a memory scrapbook to share with her mother? What were her ulterior motives with respect to her mother, father, and her own self-interest? What in the text suggests these other motives?

DIVERGENCES

Print

Carruthers, Mary. *The Book of Memory: A Study of Memory in Medieval Culture*. Cambridge: Cambridge University Press, 1990. This study discusses the elaborate spatial mnemonic system Lamy attributes to Cicero as it was used in medieval times.

Web

http://www.bookwire.com/bbr/bbr-home.html. Nicole Lamy has served as managing editor of the *Boston Book Review*.

HOMELESS
Margaret Morton

2 PHOTOGRAPHS, ORAL HISTORY

This selection revisits the topic of homelessness that was first broached in Mary Ellen Mark's photos of the Damm family in Chapter 2. It, too, is an attempt to put a human face on the issue without resorting to overwrought sentimentality. You may want to invest some time talking about (or even researching) the issue of growing homelessness in the United States, and how images and conceptions of homeless people have changed with this growth. For instance, when we think of the homeless, do we imagine stereotypical images of old winos, lovable Dickensian street urchins, or some other type of person?

In part, Morton's aim in this selection is to highlight Mr. Lee's personality, to make him stand out as an individual instead of a nameless, faceless transient. Ask your students to identify and explain the different ways in which she does this. For instance, what purpose does the picture of Mr. Lee's house (as well as Morton's naming it such) serve in telling an audience about Mr. Lee? What effect is Morton trying to achieve with her portrait of Mr. Lee, seated on a park bench with his belongings tied up in a sack at his feet? What personal details does the oral history provide that show Mr. Lee as a specific person, and how does this past explain his present existence?

Finally, you might have students consider where the artistry is located in this selection. Is it Morton's photographs alone—which take images out of their real-life context and place them in a photography book alongside other homeless dwellings—that give Mr. Lee's house an aesthetic quality? Or is there something in Mr. Lee's design itself that suggests his shelter is more than just functional? What purpose is served by the various ornamental features of the house, such as the calendars, the signs, and the teddy bear? You could discuss with your students the possible symbolic importance of these objects, as well as why Mr. Lee chooses to make his house more than basic shelter in the first place.

MESSAGE

Because Mr. Lee's house is held together with knots, it certainly does not allude to conventional architecture, nor does it conform to the cardboard-box, mound-of-blankets imagery we associate with makeshift homeless dwellings. In this house, ornamentation seems to be the main feature, in the sense that the knots themselves appear to be decorative—it's almost as if the hut proudly showcases the fact that it is a hut. Morton's depiction of Mr. Lee gives viewers a sense of a man whose desire for a home is so strong that, even without the opportunity for a proper dwelling, he built one out of trash and with no small amount of care.

METHOD

Mr. Lee's house suggests a certain ingenuity on his part. Lacking conventional building materials such as nails and boards, he resorts to found items, reinventing them so they conform to his image of a shelter. This self-supporting structure is also adorned with placards written in Chinese; moreover, the eco-friendly structure (made from recycled waste found in the immediate environment) is a marked contrast to the spacious architecture of American standards (think back to Alex MacLean's photo of the trophy house).

MEDIUM

As Morton says in the accompanying oral history, "Much like his house, Mr. Lee is soft and round and held together by knots," and both photographs are composed so that the subject is centrally, prominently placed. Again, Morton reminds the audience of Mr. Lee's ingenuity when faced with limited resources, but students may want to consider any possible negative effects of her portrayal.

ADDITIONAL WRITING TOPICS

1. If you were thrust into a situation that demanded you build shelter for yourself, but you had no money or help, how would you go about this task? Consider the types of materials that might be at your disposal as well as the design you would use.
2. As the oral history indicates, Mr. Lee's house is decorated with cardboard signs reading, "CONGRATULATIONS TO MR. LEE FOR HAVING A BIG COMPANY, HE HAS HUNDREDS OF THOUSANDS OF WORKERS, EACH WORKER GETS PAID $500 A DAY, PROSPERITY TO MR. LEE, MR. LEE THE GREAT INVENTOR." Think about Mr. Lee's intentions in making these signs (remember that they are written in Chinese). Are they humorous or ironic? Motivational? Wishful thinking? Are they intended for an audience? Explain.
3. Refer to the quote by Alan Trachtenberg in this selection. What elements about these two photographs indicate that Morton is trying to empower Mr. Lee?

DIVERGENCES

Print

Balmori, Diana, and Margaret Morton. *Transitory Gardens, Uprooted Lives*. New Haven: Yale University Press, 1993.

Eighner, Lars. "On Dumpster Diving." In *Homelessness: New England & Beyond*. Edited by Padraig O'Malley. Amherst, Mass.: University of Massachusetts Press, 1992. This satirical take on food-gathering practices of the homeless is a perennial student favorite.

Morton, Margaret. *Fragile Dwelling*. New York: Aperture, 2000. This volume features an introduction by cultural studies scholar Alan Trachtenberg.

———. *The Tunnel: The Underground Homeless of New York City*. New Haven: Yale University Press, 1995.

Thoreau, Henry David. *Walden; and, Civil Disobedience: Complete Texts with Introduction, Historical Contexts, Critical Essays*. Edited by Paul Lauter. Boston: Houghton Mifflin, 2000. This is a recent and expert critical edition.

Web

http://www.cooper.edu/art/nyphoto/. The Cooper Union School of Art sponsored this three-photographer show entitled "New York Photographs" in 1998; it features Margaret Morton along with Robert Rindley and Christine Osinsky.

http://hardpress.com/newhp/lingo/authors/morton.html. The Hardpress site includes an online photo-essay by Margaret Morton entitled "José Camacho's House," which features quotes by Camacho and Morton's pictures of his dwelling.

INVADING TERRITORY
A Story about Place

FICTION, PHOTOGRAPH

Julia Alvarez's "Neighbors" is a short vignette in which the action is restrained and the reflection on that action is highlighted. A married couple—Dominican wife and American husband—encounter a part-Haitian child in a migrant worker camp on the way to their mountain home near Manabao. After the wife unsuccessfully attempts to befriend the child, the couple drives away, the husband chiding his wife's effort. One important rhetorical device used in this story is digression; have your students locate its appearance. Then, discuss with them the importance of the digression with respect to the story as a whole—the historical allusion to the Haitian massacre of 1937, as well as the general description of the exploitive treatment of the migrant workers by the Dominicans, gives readers a context for explaining the little girl's actions as well as the behavior of those who remain hidden inside the shanty.

Because the story is so intimately related to Polibio Diaz's photograph, you might want to narrow your students' focus to just the picture, having them analyze its composition. What tone is Diaz trying to convey in this picture? Although the photo is full of bright primary colors, the close cropping, the dark interior of the hut, and the girl's curious expression (fear? shock? horror?) all seem to work against lightheartedness. In such a seemingly simple image, is there some political statement driving Diaz's camera?

CALLOUT QUESTION

This question lends itself well to an informal writing exercise where students construct alternate stories around this photograph. Although Alvarez's story plays with the relationship between language and culture (a common theme in her writing, and one the photograph nicely echoes), you might want to encourage students to read the photograph in other ways: making the girl a central figure in the narrative, imagining the events going on inside the building that led up to the girl's reaction, or describing the photographer's imagined interaction with the girl.

MESSAGE

Alvarez describes pulling "the pickup onto the narrow shoulder in front of the plywood shack," the smell of lye soap coming from inside the house, and other slight details to quietly construct a sense of place. The story's title, "Neighbors," is meant in an ironic sense—the Haitian workers are not permanent residents of their camp, nor are they friendly with the main characters in the story as we might imagine neighbors to be. The American husband is characterized by an obnoxious, know-it-all attitude toward these workers; the girl, hand clasped over mouth throughout, is practically a noncharacter in this story and more a symbol of the exploited and abused transient Haitian population.

METHOD

Alvarez's point of view in this story might have to do with her own upbringing as a Dominican child in New York City—she is able to see the scene from both perspectives, through the eyes of both cultures. That she writes about the husband in unflattering language (he "retorts" with exasperated comments that indicate his superior attitude) suggests the author's affinity toward the wife, who attempts to be compassionate despite the husband's insistence that they have nothing to do with the girl. The implied answer to the husband's final question, assuming the voice of the wife, might be that she had hoped for friendly interaction with the young girl.

MEDIUM

Like Alvarez's story, which is selective in its use of detail and imagery, Diaz's photograph has an ambiguous quality about it—it doesn't convey much information to the viewer. The viewer may not be sure exactly what is going on, but the girl's reaction indicates that it is something important, and this curiosity may have intrigued Alvarez to the point of creating an equally ambiguous story in "Neighbors."

ADDITIONAL WRITING TOPICS

1. Choosing a photograph at random, write a short story in a style similar to Alvarez's. Pay particular attention to how she constructs interaction through dialogue, as well as her sparse but productive use of adjectives.
2. Alvarez alludes in the story to a massacre of Haitian migrant workers in 1937 that left twenty thousand dead. She follows up by saying, "But it was not an important holocaust." What is the intended effect of using the word "holocaust," and for what reason do you think Alvarez references this event in the context of this story?
3. Analyze Alvarez's story in terms of how she uses color symbolically. Be sure to quote specific references to colors throughout the text.

DIVERGENCES

Print

Alvarez, Julia. *How the Garcia Girls Lost Their Accents*. Chapel Hill, N.C.: Algonquin Books of Chapel Hill, 1991.

———. *In the Time of the Butterflies*. Chapel Hill, N.C.: Algonquin Books of Chapel Hill, 1994.

———. *Yo!* Chapel Hill, N.C.: Algonquin Books of Chapel Hill, 1997.

Danticat, Edwidge. *The Farming of Bones*. New York: Abacus, 1999. This novel is set against the backdrop of the Haitian massacre of 1937.

Web

http://www.doubletakemagazine.org/. Diaz's photo and Alvarez's story were taken from *Doubletake* magazine's Web site.

http://www.thirdworldtraveler.com/Blum/DominicanRepublic_KH.html. This page presents a selection from William Blum's *Killing Hope*, a book about the 1961 assassination of Dominican dictator Rafael Trujillo and the United States' involvement in reestablishing a Dominican government. The anti-Trujillo uprising caused Alvarez's family to flee the Dominican Republic and move to New York City.

PRIVATE VERSUS PUBLIC SPACE
Nick Waplington

3 SCREEN SHOTS

Waplington's screen shots invite the opportunity to talk about the concepts of parody and satire. Have students define these terms, and ask them to list examples of parodies or satires they have read or seen before—this rich literary tradition includes such varied texts as Swift's *Gulliver's Travels*, the fake rock band Spinal Tap, or the recent Wayans Brothers film *Scary Movie*. Students can generate theories about what makes these literary forms funny and, more specifically, why Waplington's pictures are humorous.

Turning to the screen shots themselves, you might begin by asking students why Waplington has chosen not to put these up as live sites but instead to exhibit them as mounted prints. Does he feel the need to take these images outside the medium for a particular reason? Perhaps so they would not be mistaken as legitimate? Point out the features that make the sites appear authentic (a successful parody relies on a close resemblance to the source), such as the punk-rock, broken-typewriter aesthetic associated with anarchy.co.uk or the use of bulleted lists and navigation buttons on Child Prodigy.

Also, direct students to some of the more surprising aspects of these sites—the seemingly out-of-place "Lovely Latinas" banner ad; the veiled allusion to Michael Jackson's ("Jacko") pedophilia on the Child Prodigy site; the request for credit card information on both the anarchy.co.uk and Socialist International sites (organizations expected to have anticapitalist agendas); the link to romantically inclined "Chatboxes" on the Anarchy site (touting "Freedom fighters against the system need love too!"). Have students suggest other parody sites. This could lead to a larger final project in which they design their own mock sites.

MESSAGE

Waplington's fake sites all invite visitors to supply their credit card information, which is particularly funny considering the antiestablishment, anticapitalist ethos of the anarchy.co.uk and Socialist International sites. Note that, although the concepts themselves are outright suspect, the sites' realistic layouts and rigorous detail make them look quite authentic. The subversive tone in Waplington's screen shots carries a wit that doesn't exactly echo the love-filled imagery of his photographs, but it is possible to find love for society's eccentricities in the diverse sectors of culture he chooses to parody.

METHOD

Certainly, Waplington's target in his Child Prodigy screen shot is those parents who carry overly high expectations of their children—ostensibly parents from an upper-middle-class background. The banner ad, the reference to Michael Jackson, and the curiously dismal picture of a deserted roadway suggest that the business organization behind this site is probably more interested in bilking customers out of their money than in serving as agents for "child prodigies." On another level, the site may be poking fun at Prodigy, the once-powerful, though often vilified, Internet service provider. The gauche, gaudy design, as well as the exaggerated content (assuming without question the star quality of the "visitor's" child) indicate a parodic intent, though considering the flood of get-rich-quick business ventures on the Web, its effectiveness as a spoof may be limited.

MEDIUM

Waplington considers the Internet a new medium because of its balanced use of visual and textual symbols, as well as its ability to allow almost anyone to publish his or her message (something that television, film, and even print can't do to the same extent). While the Internet might give voice to subversive minorities, certain opportunistic factions would gladly "borrow" that subversive voice in order to reap the same old gains—in the case of all three of these "sites," money. The rise of the Internet has produced the largest public space ever, and Web sites are out for all to see. This virtual elimination of private space has some dire consequences; relatively small revolutionary groups such as anarchists, for example, are in danger of becoming a commodity.

ADDITIONAL WRITING TOPICS

1. Choose one of Waplington's Web site parodies and write content for it, remaining consistent with the style of humor suggested by the screen shots. You might consider writing such items as a mission statement for the site, a series of bulletin board postings, or a FAQ (Frequently Asked Questions) page.
2. In your own words, define "parody." Give examples of other parodies you have encountered, explain how they function, and compare them with the selections included here.
3. Do you think there are topics about which people should never make jokes under any circumstances? If so, identify those topics and explain why humor would be inappropriate. If not, explain how humor can be beneficial during situations when others might find it tasteless.

DIVERGENCES

Print

Nick Waplington. *The Indecisive Memento*. New York: Aperture, 1999.

———. *Living Room*. New York: Aperture, 1991.

———. *Other Edens*. New York: Aperture, 1994.

———. *The Wedding: New Pictures from the Continuing "Living Room" Series*. New York: Aperture, 1996.

Web

http://www.depauwgallery.com/Top/artists/artist12.html. This site has several of Waplington's fabricated screen shots and a short bio on the artist.

http://www.theonion.com. *The Onion*, an extremely popular satirical newspaper, presents an even more popular online version.

http://www.2600.com. This hacker site features a gallery of hacked sites, many of which have been defaced in a parodic fashion.

DEATH ROW
Benetton

COMMENT, ESSAY, 3 PHOTOGRAPHS

In order to avoid volatile classroom debate (which discussion about the death penalty might provoke), you may want to shift focus away from the morality of capital punishment. Instead, engage students in a conversation about divisive topics in general, and ask them what makes certain topics (e.g., abortion, gun control, homosexual marriage) impossible to agree upon. As a research activity, you could have students look at the language involved with arguments on both sides of issues such as those listed above in order to pinpoint the specific religious or ideological positions that keep these arguments from reaching a compromise.

Next, have students work through the Benetton advertising campaign photos, focusing on how the three images together make a single political statement. What sort of mood is photographer Oliviero Toscani trying to conjure in each photo? Is the point to make each convict a sympathetic figure; is there some attempt to manipulate our perception?

Barbara Ehrenreich's "Dirty Laundry" offers students examples of several well-managed rhetorical strategies. As an argumentative essay, it is particularly adept at not settling on one side or the other of a split issue; she is able to intelligently critique Benetton's use of political agendas to further its image as a socially compassionate clothing company while simultaneously admitting to the genuine, compelling power of the ad campaign. By opting for a "both/and" approach to an "either/or" problem, Ehrenreich gives students a method by which they can approach difficult topics from positions that haven't already been established through long, tired discourse. You can discuss with your students the way that Ehrenreich complicates her narrative voice in this essay as well, sometimes admitting to very personal, human feelings about the death penalty and UCB's magazine, and at other times making a transition into analytical prose. The essay makes all three classical rhetorical appeals—to logic, to emotion, and to the writer's character—which you could direct students to locate and analyze in terms of their effectiveness. Also, the essay makes use of a well-used structure—the notion of "on one hand . . . on the other hand . . ."—but makes it more sophisticated. For example, in paragraph 6, Ehrenreich gives new life to this argumentative strategy when she writes, "On the other hand—well, the other hand reaches out for substance, for sincerity, and comes back soiled." A related rhetorical exercise might involve having students find ways of livening up their own examples of clichéd language.

MESSAGE

Ehrenreich feels that Benetton's ad campaign is valuable in raising social awareness about the inhumane nature of capital punishment in the United States (as well as much of the rest of the world's disagreement with the practice), but she still cannot shake the fact that this message is delivered in the form of an advertising campaign for a clothing manufacturer. In the end, she doesn't see the issue in either/or terms, but accepts both "hands" of her argument. Her call to throw the inmates to the lions is meant to provoke the reader; she isn't serious but she does want to point out that she finds our current capital punishment methods cruel despite our attempts to sanitize or civilize them.

METHOD

Ehrenreich is drawn to the expressions of these condemned souls, especially their eyes, which communicate the kind of humanity that printed words are incapable of conveying to her. You might want to focus on Ehrenreich's descriptions of these photographs, asking students to compare their own descriptions to the author's.

MEDIUM

Paragraph 7 of "Dirty Laundry" offers a succinct analysis of this advertising strategy and how it works. "Branding," Ehrenreich suggests, is a strategy that involves more than just making the argument that the product itself is worth buying; an attitude or a political sensibility is associated with that product so that consumers feel they take on a certain identity when buying it. That this strategy has been used for several decades indicates at least that the advertisers themselves feel it is effective. Have students conduct a brainstorming session on ad campaigns other than the ones Ehrenreich mentions: Nike, Marlboro, and Apple.

ADDITIONAL WRITING TOPICS

1. Which of these three photographs did you have the strongest reaction to when you first saw them? Characterize that reaction: sympathy, anger, frustration, or sadness? Which specific aspects of the photos evoked this reaction?
2. What message do you think the photographs are ultimately successful in conveying—pro- or anti-capital punishment? What message were they intended to convey? Where else in popular culture (films, television, newspapers, radio) can you find support for or arguments against the death penalty?
3. After reading Barbara Ehrenreich's essay "Dirty Laundry," write a letter to her in which you respond to the piece. You might wish to focus on the argument itself (whether you agree or disagree, or whether it made you think differently about the topic), her style of writing (her lively use of tired metaphors, or her witty tone), or even share a related personal experience in order to support or refute her position.

DIVERGENCES

Print

Toscani, Oliviero. *Current Biography*. (Sept. 1998): 55ff. An article about Benetton's controversial photographer, as well as historical information on the campaigns themselves, which have set out to tackle topics ranging from racism to AIDS.

Web

http://www.benetton.com. The official United Colors of Benetton Web site contains information about their various advertising campaigns as well as an online catalog.

http://www.benetton.com/colors. The United Colors of Benetton also has an online magazine, *Colors*.

Audio/Visual

Dead Man Walking. Directed by Tim Robbins. Starring Sean Penn, Susan Sarandon. 122 minutes, rated R. MGM/UA Studios, 1995. Videocassette. This acclaimed, Oscar-winning film is based on the true story of Louisiana nun Sister Helen Prejean's attempt to save a condemned killer from execution.

4 Making History

For today's students, history, as a discipline, is a paradox. On the one hand, students learn about the midnight ride of Paul Revere, the French Revolution, Eli Whitney and the cotton gin, and other topics seemingly disconnected from modern life. On the other hand, as the introduction to this chapter notes, the film industry thrives on blockbuster movies portraying various historical figures and incidents, by first stripping them of any content or context that may be unentertaining and then supplying attractive men and women, enormous sets, and an aggressive media campaign. How do we explain many students' lack of interest in traditional history yet the massive success of the movie *Titanic*?

One explanation may lie in the question asked in the introduction: "Since the past has vanished, how *do* we accurately recover it?" When we read a history textbook, are we not really just reading an author's rendition of the facts as we know them? What constitutes a historically "true" account? History is essentially an abstraction, and we attempt to recover it and make it knowable by creating a narrative to relate the facts in a concrete way. For example, David McCullough's popular biography *John Adams*—which recently made the *New York Times* best-seller list and sold over a million copies—is considered by some to be soft on the facts, an easily digestible history written for the masses. This does not suggest that the public dislikes history but rather that students are not interested in being fed the same version of history their parents were—they need a narrative that relies less on facts and more on the human feel of a time and place.

You may want to begin discussion of this chapter by asking students what they know of the real *Titanic*, the actual presidency of John F. Kennedy, or another example from the chapter. Then ask them how they came to know this information. Did they get it from a textbook? A magazine? A movie? Which source is the most reliable? Have students recount their own histories as well—ask them about their first memory or the color of the walls in the house they grew up in. How do they remember this? Many will cite photographs as the main sources of memory. Does this mean that they may not remember things as they really were? Is that even possible? This exercise could then lead into a discussion of the first selection, "Solemates," and how objects from a certain time period sometimes recall that time more effectively than human memory does.

SOLEMATES
The Century in Shoes

3 SCREEN SHOTS

Because it is so unconventional, this selection provides for a good introduction to the theme "making history." Ask students to give their own definitions of "history": very likely, they will echo conventional conceptions of history—dusty books about kings, wars, and political affairs. After establishing a tentative definition, have students explore the 4th Revolution's Web site "Solemates" in small groups and return to the question of defining history. Can "Solemates" be seen as a legitimate representation of history? If so, how does it alter the original definition? You could explore this line of questioning with regard to the site's subject matter, the medium of the Internet, and the fact that it is an advertisement designed to showcase 4th Revolution's services.

You could also ask students to note how developments in shoe design over the decades mirror broader cultural changes, as well as how the site works to make these claims (see Callout Question below). Ask students about the validity of looking at shoes—a seemingly inconsequential subject—in terms of "reading" a particular culture. Can we develop new perspectives for studying an era by examining ordinary objects and events?

By now, you may have had your class visit several Web sites, and a comparison of how different sites categorize and organize their content could prove beneficial. Examine how site architecture accomplishes the goals of the sites' authors—Physicians Against Land Mines and Negative Population Growth aim to raise awareness as well as solicit financial support, maganda.org seeks to share personal information with an audience to establish community, and 4th Revolution hopes to attract new advertising clients. This could be a useful exercise for many reasons, especially if your students produce Web-based projects of their own, or if you want to help them understand through analogy how different organizational patterns can be used for different effects in their own writing.

CALLOUT QUESTION

While it is certainly easy to see how changing shoe designs over the decades reflect our culture's changing aesthetic taste and our sense of fashion, it is not so apparent how shoes can comment on more general cultural changes. By juxtaposing shoes and suggestive quotes, images, advertisements, and other information to paint a singular picture of each decade, the "Solemates" site makes the argument that, for a given era, shoe design is a good barometer of its zeitgeist—the cultural spirit of a particular time. For example, the skimpy, colorful open shoe designs of the 1920s are meant to suggest a partying, carefree era, a suggestion echoed in the image of the flapper as well as the adjectives used in the text ("risqué," "tumultuous," "precocious"). Similarly, the use of synthetic materials in shoe designs of the 1970s reflects that decade's growing dependence on high technology and its preoccupation with the promise of science.

MESSAGE

Throughout the "Solemates" site, the image of the shoe is the unifying element, the metaphor intended to let the user into the mind-set of each decade. The pages are designed so that the images of the shoes are placed in the context of literary quotes, advertisements, a short historical description, and a large picture of a person from each decade. Neither the visual nor the verbal information is more important; both work together to provide the whole picture.

METHOD

Though some may find the site a bit too "click-heavy" (there are many separate pages), the navigational cues are immediately intuitive. For example, the dial (accompanied by a decade-appropriate telephone graphic) is a familiar metaphor borrowed from an established technology, and the navigation bars on the top and bottom margins of each page are by now equally familiar. Also, the brief textual material is broken up over several pages, which, as some Web designers argue, improves legibility. As a method for presenting history, this Web site is accessible and easily digestible, but some might argue that it is not thorough enough to offer readers a "true" perspective of the historical eras it covers.

MEDIUM

"Solemates" showcases 4th Revolution's sense of inventiveness and creativity—shoes certainly make for an unorthodox historical and Web-based subject. By showing prospective clients how the firm might create an entertaining, engaging site out of odd content (rather than simply bragging about its skills in general), 4th Revolution is able to suggest to them how their own content might be "realized."

ADDITIONAL WRITING TOPICS

1. Describe your favorite pair of shoes (or, alternately, your favorite piece of clothing). In addition to providing a detailed physical description, recall how you originally got the shoes (or clothes) as well as any special memories you associate with them.
2. "Solemates" raises the question of which medium in our current era is most appropriate for documenting history. With a partner, discuss the advantages and disadvantages of cataloging historical evidence relative to various media, and propose an answer to this question.
3. Traditional approaches to history have long followed the model established by nineteenth-century German historian Leopold von Ranke, whose aim was to depict events "as they happened." Consequently, history was about monumental events—battles, conquests, and not much else. As a history, "Solemates" departs from this approach by making an otherwise ordinary topic the central focus. Construct your own list of topics similar to shoes that would make for good objects of historical study. Explain why in a short follow-up paragraph for each topic.

DIVERGENCES

Print

Lawlor, Laurie. *Where Will This Shoe Take You?: A Walk Through the History of Footwear*. New York: Walker & Co., 1996. This children's book on the history of shoes can be used to explore the implications of mixing textual and graphical elements in print.

Warhol, Andy. *Shoes, Shoes, Shoes*. New York: Bulfinch Press, 1997. This is a collection of the postmodern artist's take on the subject matter at hand.

Web

http://www.4threvolution.com. The 4th Revolution's portfolio site contains background information on the marketing communications company that developed "Solemates." The firm's clients include Adobe Systems, Amtrak, Hewlett-Packard, Lucent Technologies, and Wells Fargo. Explore one or more of their corporate Web sites to see how 4th Revolution's design projects take final shape.

Audio/Visual

The Red Violin. Directed by François Girard. Starring Samuel Jackson. 143 minutes, rated R. Rhombus Media, 1998. Videocassette. This fictional film follows a well-traveled violin as it passes to owners from various countries and cultures over three hundred years. It does for the violin what "Solemates" does for shoes.

MONTAGE
Danny Lyon

4 PHOTO-MONTAGES

Danny Lyon's montage art reflects his take on personal history. In Lyon's work, history functions as a composite of several images (a montage) that come together to make a single statement, a vague collection of memories that unlock countless associations. The montages included in the text invite an exploration of the different uses of photography, which has enjoyed a rich history as a multipurpose medium. You could discuss the changing role of photography through the years by connecting Lyons to other photographers you have studied—Sally Mann, Cindy Sherman, William Wegman, Weegee. In addition to Lyon's artistic use of the technology, you could also discuss how photography functions in such spheres as journalism, personal documentary, social commentary, advertising, and even political propaganda. In the pursuit of a course-long theme dedicated to understanding the roles and capacities of different media, this exercise or conversation could easily be extended to account for the various functions of film, radio, television, print, painting, digital technologies, and so on.

Ask students to define Lyon's style and describe how it has changed over his career. You might have them note that the pictures in *Ernst* were not taken by Lyon himself, which complicates our understanding of his artistic role in the piece. Does his assemblage, along with the addition of a caption and title, make the piece truly his own? Additionally, does the montage featuring his family members differ in tone from the montage of his friends and classmates in *First Photographs, 1960*? Whom does he depict in a more intimate light—his family or his friends? Furthermore, how do we know these particular pictures are from 1960? Finally, how does the inclusion of Ferris wheel photos and carnival imagery indicate a change in Lyon's style?

As a multimedia exercise, have students form groups and assemble stock photographs into a series of their own montages, providing short narrative glosses modeled after Lyon's. An exercise such as this would be designed to foster students' familiarity with the medium; it could also help strengthen their understanding of how the interaction of word and image create meaning.

MESSAGE

The captions to the four photo-montages serve to augment the reader's understanding. They not only identify the people in each photograph but also tell the stories around the pictures. For example, the caption to *Raphe at 17, 1992*, refers to a conversation that is impossible to capture with a camera, but certainly adds a level of characterization to the figure we see in the central picture. One possible class exercise might involve talking about how the montages might be read differently if there were no accompanying captions.

METHOD

Lyon's mix of black-and-white and color images emphasizes the fact that these pieces are montages—collections of separate and possibly unrelated photographs. In terms of reconstructing memory, Lyon might be implying that memory is a construct made up of several smaller, unrelated "snapshots" that congeal into a single impression. The method of composition—a central photograph framed by other photos—guides the viewer's eye to the central element, which suggests the importance of that central picture in forming the montage's theme.

MEDIUM

Montage is a technique usually associated with film editing, and it works to form meaning by placing two disparate images together: when a shot of an actor talking is spliced with a shot of another actor looking into the camera, the audience understands the two characters to be having a conversation. A variety of techniques could be used to translate Lyon's montages into documentary format, though the inclusion of audio elements and movement would be primary considerations. Consistently, Lyon's pieces indicate a strong relationship to family, friends, and a personal curiosity for history.

ADDITIONAL WRITING TOPICS

1. In reference to his father's photos, Danny Lyon wonders, "How can photographs affect us so, transporting us through time, across continents, and beyond wars? How can they preserve what is gone forever?" Think of a picture in your life to which you have a particularly strong emotional attachment. Describe both the picture itself and the associations you have for it.
2. Explain the quote by art historian Rod Slemmons included in this selection. In what way do Lyon's montages "witness" events, according to Slemmons? How are they "live"? Do you agree or disagree with Slemmons's assessment of Lyon's work?
3. In a group, respond to one of Lyon's montages included here. Consider separately such details as how your eye moves over the page or how you react to shifts from color to black-and-white images. Compare your impressions with those of the rest of the group. Were they the same? Why or why not?

DIVERGENCES

Print

Lyon, Danny. *The Bikeriders*. New York: Macmillan, 1968.

———. *Conversations with the Dead*. New York: Holt, Rinehart and Winston, 1971.

———. *The Destruction of Lower Manhattan*. New York: Macmillan, 1969.

———. *I Like to Eat Right on the Dirt: A Child's Journey Back in Space and Time*. New York: Bleak Beauty Books, 1989.

———. *Knave of Hearts*. Santa Fe, N.M.: Twin Palms Press, 1999.

———. *Pictures from the New World*. New York: Aperture, 1981.

Web

http://faculty.washington.edu/rods/art_lyon.html. This URL takes you directly to Rod Slemmons's essay on Lyon.

http://www.kodak.com/US/en/corp/aboutKodak/kodakHistory/kodakHistory.shtml. Kodak's extremely thorough history of its technological, cultural, and corporate milestones offers a nice parallel to Lyon's work by explaining the changing medium of photography through the years.

http://www.nara.gov/exhall/picturing_the_century/portfolios/port_lyon.html. The National Archives' exhibit "Picturing the Century" includes a portfolio of Lyon's work.

Audio/Visual

The Abandoned Children. Directed by Danny Lyon. 63 minutes, not rated. 1974. Videocassette. Lyon's documentary about street children in Colombia may be compared to his still photography.

Little Boy. Directed by Danny Lyon. 52 minutes, not rated. 1976. Videocassette.

El Otro Lado (*The Other Side*). Directed by Danny Lyon. 60 minutes, not rated. 1978. Videocassette. These and more of Lyon's social documentaries can be located through http://www.buyindies.com.

HOLLYWOOD
Moving History

INTERVIEW, 6 FILM STILLS, REVIEW

As an entry point into this selection, you may want to take advantage of the likelihood that many of your students have already seen James Cameron's *Titanic*. You might begin by asking about their reactions to the film. Then, ask them to compare their own reactions with those stated in Michael Wilmington's review—did they notice things about the film that Wilmington neglects, and vice versa? You could follow this comparison with a focus on the rhetorical setting of Wilmington's writing. Instead of writing in a forum where he can simply voice his opinion of the film, he must write a column that will please those who like the film and simultaneously convince the skeptics in his audience that the film is worth seeing. As a result, his review is highly complimentary, yet it also acknowledges the movie's faults. He calls it an amazing spectacle and a prime example of modern filmmaking, even if it is unbelievable and almost too extravagant (though he is careful to point out that movies are allowed to commit these crimes). Ask students to find passages from the text that indicate how Wilmington manages to give the film careful praise.

Next, it would be a good idea to direct discussion to the more general topic of defining the relationship between film and history. The Oliver Stone/Mark Carnes interview is a good background text because it sums up the issue from both sides and both agendas—that of the filmmaker who wished to have complete freedom in his art, and that of the historian who demands that the "truth" be told. Historian Mark Carnes expresses concern over Hollywood's occasional disregard for known facts when making films on historical subjects, and he laments the oversimplification of life's complexities that results from filmmakers' adhering to dramatic structures. Director Oliver Stone, in contrast, embraces the artistic freedom inherent in filmmaking; he feels that movies can offer interpretations of historical events that are just as valid as those put forth by historians. Also, he argues, because the "facts" that historians play with are often suspicious, it becomes the job of the filmmaker to document a "truth" that engages the viewers' emotions.

In dealing with the film stills included with this selection, you might want to discuss the truth/fiction divide. How do we "know" that the pictures used in the film *JFK* are real, and that the event depicted actually happened? How does our reaction to these pictures differ from our reaction to the others, which we know to be fabrications and embellishments of historic events? Finally, is there a difference between how we read the *Nixon* stills, which show actors depicting actual people performing actual events, and how we read the entirely fictitious (and therefore unrealistic) pictures of *Titanic*'s Jack and Rose?

MESSAGE

Michael Wilmington's main point of contention concerns the plausibility of *Titanic*'s love-story plot (and its purportedly integral role in causing the disaster), but he is perfectly willing to suspend his sense of logic in favor of experiencing the satisfying emotional release and visual pleasures the film offers him as a viewer. Thus, he could be seen as going "soft" on the film. One might easily speculate that this viewpoint would cause Mark Carnes a bit of uneasiness—as a historian, Carnes would value authenticity and factual reverence over a good story. Conversely, Oliver Stone, who values artistic license and creativity over accepted notions of reality, would likely support Wilmington's conclusions.

MEDIUM

Wilmington admits that because *Titanic* is a film, it does not have to adhere to our sense of historical truth—he says, "Plausibility isn't important here." What is important, though, is the film's ability to impress spectators, to retell an actual historical event in terms of universal emotions. Some would argue that this is the function of great art and literature as well, and it is not necessarily media specific. Carnes and Stone debate the ability of film to properly convey the many intricate elements that surround a historical event or period—that kind of interpretive work, Carnes suggests in his last statement, is done better in print than on film. Stone disagrees, of course, claiming that the dramatic nature of film gives us other, more imaginative ways of looking at events that traditional historians overlook.

METHOD

Carnes might approve of documentary film as a responsible medium for history, but he more likely feels that writing offers the kinds of analytical interpretations necessary for making sense of the past. Director James Cameron's goal in constructing *Titanic*, however, was to create a grand spectacle, the kind of media event that would appeal to a broad range of people (and this includes commercial considerations as well). The romance plot of *Titanic*, therefore, functions as a device that allows the audience to identify with the subject matter of the film; the average moviegoer probably understands the tragedy associated with the lost love of two beautiful stars more than with the drowning of fifteen hundred anonymous passengers. Had Oliver Stone directed *Titanic*, the movie might have had a darker, more insidious tone, considering Stone's noted interest in conspiracy theories.

ADDITIONAL WRITING TOPICS

1. Does Hollywood have the responsibility to depict events in a historically accurate light or not? Defend your answer, citing quotes from the interview between Mark Carnes and Oliver Stone.
2. After examining the style and structure of Michael Wilmington's review, choose a recent film that either entertained or displeased you greatly and write a review of your own.
3. Theorists of history have long maintained that the problem with the past is that it is irretrievable—we can never fully re-experience events that have already occurred, and every attempt to document or capture such events is imperfect, so we'll never know the "true" past. In your opinion, which medium has the greatest capacity for capturing events as they actually happened? Explain your choice.

DIVERGENCES

Print

Carnes, Mark C. "Past Imperfect: History According to the Movies." *Cineaste* 22, no. 4 (Fall 1996): 33–38. This article contains complete text of Carnes's interview with film director Oliver Stone.

———, ed. *Novel History: Historians and Novelists Confront America's Past (and Each Other)*. New York: Simon & Schuster, 2001. This recent collection explores the role of literature in shaping our historical consciousness.

———, ed. *Past Imperfect: History According to the Movies*. New York: Henry Holt, 1995. This is a collection of articles by various authors on Hollywood's approach to history.

Web

http://www.historyinfilm.com/. This is an excellent educational resource that catalogs and annotates films about historical subjects and periods, with related facts, clips, and other quotable material.

Audio/Visual

JFK. Directed by Oliver Stone. Starring Kevin Costner. 189 minutes, rated R. Warner Bros., 1992. Videocassette.

The Patriot. Directed by Roland Emmerich. Starring Mel Gibson. 165 minutes, rated R. Warner, 2000. DVD.

Titanic. Directed by James Cameron. Starring Leonardo DiCaprio, Kate Winslet. 192 minutes, rated PG-13. Paramount, 1998. Videocassette.

CHICAGO
Wayne F. Miller

3 PHOTOGRAPHS, ESSAY, POEM

The conventional image of Chicago's South Side in the 1950s might be criticized as somewhat romanticized: a lively cultural hotbed that helped give shape to a particularly urban style of blues and jazz and other art forms, akin to the earlier Harlem Renaissance in New York City. To an extent, this depiction is true, but it doesn't give us the full story surrounding the growing black population of midcentury Chicago. Wayne F. Miller's photographs offer us a different view of this emerging world, one that doesn't gloss over the poverty, the segregation, and the occasional despair felt by these urban settlers. From the crowded kitchenette to the tavern exterior to the rabbit vendor's display, Miller's photographs emphasize the hardships of this new life by putting the viewer in the midst of the scene. We, like Miller himself, are positioned so that we stand on the street beside the young girls, we wait to buy rabbits, we feel the claustrophobia of an overcrowded one-room apartment.

Certainly, this is the opinion shared by essayist Gordon Parks, who applauds Miller's work for telling it like it is. Have students focus on the language Parks uses to describe his reaction to Miller's work—his figures are particularly violent and bleak, much like the very lifestyle he remembers. In the first paragraph alone, Parks mentions that Miller's images not only sliced through him "like a razor" but also "murdered" time for him. After charting similar language throughout the piece, question students about the rhetorical effectiveness of using embittered language such as this, especially in a piece of writing that is meant to commend the photography of Miller.

This selection raises questions about how different media function differently to convey information—namely, within the medium of print, how do different genres or forms of writing express ideas differently? Generally, Parks's essay and Gwendolyn Brooks's poem "Kitchenette Building" both strive to express the frustration associated with the South Side lifestyle, but they go about this task in specific ways. Whereas Parks's tone is angry in places, Brooks's is almost one of quiet defeat. Parks's violent metaphors contrast well with Brooks's use of food, color, and scent imagery. Also, the use of short stanzas and fragments in Brooks's poem could be read as a more open-ended interpretation of the environment than Parks's authoritative, declarative paragraphs. Keep these considerations in mind when asking students to analyze each piece.

MESSAGE

Rabbits for Sale, a curious scene of dead rabbits strung along a line above a city sidewalk, signifies an emerging lower class, newly arrived from the rural South and not yet fully assimilated into an urban culture. The urban environment is evident from the city streets and tenements surrounding the rabbits. You may want to have students consider not only the class-based but also the race-based implications of this image.

MEDIUM

As published art, Miller's pictures reach masses of people who ordinarily would not see these sights. In a real sense, his photos "speak" to viewers who might otherwise treat the subjects as invisible. In *Two Girls Waiting Outside a Tavern*, we see Miller's take on the relationship between drinking and poverty, a relationship that, sadly, sometimes contends with that of family (if we make the assumption that the girls are waiting for a relative, such as their father).

METHOD

Both Brooks's "Kitchenette Building" and Miller's *One-Room Kitchenette* capture the busy, hectic pace of poor urban life. It is precisely this chaos that gives birth to the dreaming Parks mentions in "Speaking for the Past"—so as to provide quiet escape from the real world. However, the problem with dreaming, as Brooks suggests, is that there is no time to do it properly—just as the speaker in her poem begins to contemplate the possibility of dreaming, reality intrudes: "Since Number Five is out of the bathroom now, / We think of lukewarm water, hope to get in it." In a sense, Brooks's poem itself is a dream—the contemplations of the poetic mind set down on paper. Miller's photographs can also be seen as capturing the dreams of an entire community of people who have come to the city in order to improve their quality of life.

ADDITIONAL WRITING TOPICS

1. Choose a partner. Separately, brainstorm a list of all the images, ideas, movies, events, people, memories, and anything else you can think of that you associate with the word "city." Compare your list with your partner's. What similarities exist, and what is different?
2. Choose one of Wayne F. Miller's photographs and compose a narrative around the events depicted in it. Alternately, compose a narrative combining the events in all three photographs included here.
3. Spend some time researching mid-twentieth-century Chicago (you might start by visiting the Web site for the Chicago Historical Society at http://www.chicagohs.org/). In what ways does the historical treatment of Chicago in the source(s) you found differ from the one offered by Miller's photography and Brooks's and Parks's writings? How are they similar?

DIVERGENCES

Print

Brooks, Gwendolyn. *Selected Poems*. San Francisco: HarperCollins, 1999.

Miller, Wayne F. *Chicago's South Side: 1946–1948*. Berkeley: University of California Press, 2000.

Web

http://www.chicagohs.org/. The Web site for the Chicago Historical Society, mentioned in the third writing topic above, may provide useful background material for this selection.

Audio/Visual

A Raisin in the Sun. Directed by Daniel Petrie. Starring Sidney Poitier. 128 minutes. Columbia Pictures Corporation, 1961 (released on video, 1987). Videocassette. This is a film adaptation of Lorraine Hansberry's award-winning play about a struggling black family living on Chicago's South Side and the impact of unexpected monetary gains. Each family member sees the bequest as the means of realizing dreams and of escape from the frustrations of poverty. The title alludes to a well-known poem by African American writer Langston Hughes.

What America Would Be Like without Blacks
Ralph Ellison

ESSAY

To begin thinking about Ralph Ellison's provocative essay, you could mediate a brainstorming session that centers on his title, having students substitute the word "blacks" with as many other concepts as they can think of that help contribute to the concept of America. This filling-in-the-blank exercise could extend beyond ethnic groups to include individuals, events, inventions, ideas, works of art, and so forth. Putting together a wide and varied list would help flesh out one of Ellison's main points: the interrelatedness of the components that construct the powerful idea of America.

Another ideological concept Ellison's essay brings into question is that of democracy, and especially the role of blacks in maintaining it. In his last sentence, Ellison mentions how this system of government often works at cross-purposes: "The nation could not survive being deprived of [blacks'] presence because, by the irony implicit in the dynamics of American democracy, they symbolize both its most stringent testing and the possibility of its greatest human freedom." Here, Ellison's point is that blacks have historically safeguarded the notion of democracy by being in a position that shows its limits (in the institution of slavery or economic oppression, for example) as well as its potential (as seen in the triumphs of the civil rights movement of the 1960s). You might lead a discussion by asking students to offer their own personal definitions of democracy, drawing attention to the paradox Ellison lays out.

Finally, you could encourage further research in the area of African American contributions to American culture. Certainly, providing a catalog of such direct accomplishments is not the purpose of Ellison's essay, though he does mention, in passing, the areas of popular culture influenced by blacks and alludes briefly to how African linguistics contributed to the evolution of American English. Having students find out for themselves the details behind these contributions would add another facet to Ellison's argument; beyond the state-approved narratives about Martin Luther King Jr. and George Washington Carver that many of us are taught, many of these legitimately historical moments are not especially well known in our culture even today.

MESSAGE

Without question, Ellison's use of the term "fantasy" is negative, and he entertains the *thought* of an America without blacks in order to disrupt or spoil the fantasy by forming an argument against it. Ellison's main argument is that America is irreducibly black; the American nation has been shaped by the black experience both good and bad, and to think it could be otherwise would be a denial of reality.

METHOD

Jazz, often described as the one truly American art form, is the result of decades of black innovation. Like the musical form itself, the African American influence on shaping American culture has been improvisational—bending rules and straying from the dominant (white) rule-bound institutions. The term "jazz-shaped," combining both sound and space, suggests a free-flowing, nonstructured conception of these cultural developments, and a "tragic-comic" approach to life. For example, whereas classical music entails regimented training and a well-defined sense of good and bad, jazz often values experimentation and embraces nonstandard approaches to generating tones.

MEDIUM

Ellison's writing style, which blends academic prose with the vernacular, parallels the broader goal of his essay, that is, to show how intimately related black and white "cultures" are in America. One such example can be found at the end of a passage that directly refers to the African influence on the English language: "So there is a *de'z* and *do'z* of slave speech sounding beneath our most polished Harvard accents, and if there is such a thing as a Yale accent, there is a Negro wail in it—doubtlessly introduced there by Old Yalie John C. Calhoun, who probably got it from his mammy."

ADDITIONAL WRITING TOPICS

1. Refer to the anecdote Ellison relates regarding Abraham Lincoln's support of Negro colonization. Were you previously aware of this bit of historical information? Does it challenge or complicate the popular image of President Lincoln? Why do you think this story is either normally downplayed or seldom mentioned?
2. Ellison states that, on the level of popular culture, "the melting pot did indeed melt, creating such deceptive metamorphoses and blending of identities, values, and lifestyles that most American whites are culturally part Negro American without even realizing it." What does he mean? Give some examples from popular culture today that have been explicitly or implicitly shaped by African American influences.
3. The headnote to this selection mentions the historian's practice of imagining counterfactuals, or scenarios of what might have happened had an inciting event not taken place. Write a counterfactual essay on a subject of your own choosing; speculate how history might have been different if a significant event had never happened. Try to be logical in constructing your fictitious narrative.

DIVERGENCES

Print

Ellison, Ralph. *The Collected Essays of Ralph Ellison*. New York: Modern Library, 1995. This collection includes a preface by Saul Bellow.

———. *Flying Home and Other Stories*. Edited by John F. Callahan. New York: Random House, 1996. This is a recent collection of Ellison's short fiction.

———. *Invisible Man*. New York: Vintage International, 1995.

———. *Shadow and Act*. New York: Random House, 1964. This collection of Ellison's nonfiction writing includes critical essays on politics, music, literature, and culture spanning two decades.

Faulkner, William. "The Bear." *The Faulkner Reader: Selections from the Works of William Faulkner*. New York: Random House, 1954. Ralph Ellison's essay refers to the exploits of the freed slave (and nuisance to white America) Percival Brownlee in Faulkner's short story.

Twain, Mark. *The Adventures of Huckleberry Finn.* 1884. Reprint, New York: Penguin USA, 1986. Twain's novel can be used in conjunction with Ellison's discussion of how Twain wonderfully captured the slave dialect.

Web

http://www.english.upenn.edu/~afilreis/50s/ellison-main.html. Professor Alan Filreis of the University of Pennsylvania maintains these pages of critical commentary on Ralph Ellison's *Invisible Man.*

http://www.freedomship.com/. The Freedom Ship Project is a rather bizarre contemporary version of gentrification—in this case, the wealthy are invited to escape a corrupted America to go live on a city at sea.

Audio/Visual

Jazz. Directed by Ken Burns. 19+ hours, not rated. PBS, 2001. Videocassette, DVD. Various segments of this highly acclaimed Ken Burns documentary can be used to emphasize what Ralph Ellison terms America's "jazz-shaped" culture.

JOIN
Propaganda and Protest

ESSAY, 3 POSTERS, SCREEN SHOT

Pat Conroy's essay "My Heart's Content" can be used to demonstrate to students a variety of unconventional techniques that they could take into their own writing. For instance, rather than adhering to a chronologically linear structure, the essay uses devices like flashback and parallel narrative, techniques that allow for critical reflection in ways that straight narrative does not. Flashback gives Conroy the benefit of a more mature sense of hindsight, and parallel narrative allows him to place his activist past alongside his friend's horrific ordeals and thus amplify his regret for the naive, youthful decisions he made. Also, you could discuss how Conroy creates his persona here. Even though he writes from his own point of view, he does not choose to make himself the "hero" of his narrative—in fact, his self-portrayal is not at all flattering. Pointing out that first-person narratives do not always require the writer to assume the function of protagonist can help students see potential for experimenting with how they work themselves into their writing.

As you turn to the U.S. Army posters and screen shot, you could ask your students to supply their own definition of "propaganda," a word that traditionally has negative connotations. You might distinguish between propaganda and advertising, noting that these ads are not designed to persuade consumers to buy a specific product but rather to take a specific action, that is, to enlist. And that is just the explicit goal—like many commercial ads, these recruitment ads try to project a certain image. To do so, they use an ideology that draws on the core belief system of the average American. You might spend some time outlining this belief system by listing terms, concepts, and ideas that the "average American" holds in esteem.

This activity should also lead into an analysis of the various images and slogans the U.S. Army has used over the years. What makes them effective as recruitment tools and as "image restorers"? Spend some time dealing with each poster and screen shot here, comparing the early stylized imagery to the later realistic pictures, the earlier short and direct slogan to the later text-heavy designs. What cultural shifts might account for the move from the authoritative, duty-bound Uncle Sam poster to the recent ads that promise personal gains and individual freedoms? How has the relationship between the U.S. Army and the pronoun "you" changed from "I Want You" to "America's Army Wants to Join You"? And what does the move from the medium of the poster to those of television and Web sites say about the type of audience these campaigns are meant to reach?

CALLOUT QUESTION

Since students may have strong opinions about what constitutes heroism and patriotism, you might present this question by taking a poll of those who feel Pat Conroy's actions during the Vietnam War were commendable and those who feel they were deplorable. Afterward, have these two groups work together to construct a debate, delivered orally or in writing, around the issue.

MESSAGE

Visiting the *Vietnam Veterans Memorial* did not move Pat Conroy in the same way that his encounter with Al did. His epiphany comes about in part because thinking about a flesh-and-blood human being's involvement in the conflict Conroy opposed forces him into a position where he empathizes with the other side. Of course, the lessening of antiwar sentiments over the years, in addition to the U.S. Army's new individualized image, probably helped soften Conroy's idealistic opinion because he can no longer think of the army as an inhuman mass of troops.

METHOD

Conroy's epiphany is not one that brings comfort or jubilation (it is not a "Eureka!" kind of moment). Instead, Conroy's self-understanding is mixed with profound regret—not at having the beliefs that he did at the time, but at lacking the ability to sympathize with those people on the other side, the war supporters and reluctant soldiers. In short, his youthful idealism kept him from being able to see issues in shades of gray. That Conroy spent much of his life after the Vietnam War ended doggedly researching totalitarian regimes suggests a need to prove his convictions right, especially to himself.

MEDIUM

Like film, Conroy's essay plays with techniques like flashback—where he asks Al for his story and the point of view slips into the anonymous third person—and parallel narratives—Al's story is interspersed with moments of Conroy's life that happened at the same time. An essay like "My Heart's Content" would probably not make a good recruitment tool because it is not propagandistic enough. Whereas propaganda relies on definite ideological messages, Conroy's essay neither endorses nor condemns the idea of serving in the armed forces.

ADDITIONAL WRITING TOPICS

1. Have you ever acted according to a principle that you believed was right at the time, only to doubt or regret your actions later? Elaborate on this episode in a reflective narrative in which you explain your feelings during and after.
2. Spend some time exploring the U.S. Army's recruitment Web site at http://www.goarmy.com. How does this site use different rhetorical approaches to "sell" the army's message to different groups? For instance, is the message to men different from the message to women? How are people of color included in this pitch?
3. Compare and contrast the American recruitment posters included in this selection with propaganda art from such countries as the former Soviet Union, post-revolution Cuba, or Nazi-era Germany. Do you notice any similarities or differences in the visual styles of these posters? In the various slogans used?

DIVERGENCES

Print

Conroy, Pat. *Beach Music*. New York: N. A. Talese, 1995.

———. *The Great Santini*. Boston: Houghton Mifflin, 1976.

———. *The Lords of Discipline*. Boston: Houghton Mifflin, 1980.

———. *My Losing Season: A Point Guard's Way of Knowledge*. Forthcoming.

———. *The Prince of Tides*. Boston: Houghton Mifflin, 1986.

———. *The Water Is Wide*. Boston: Houghton Mifflin, 1972.

Teitelbaum, Matthew, ed. *Montage and Modern Life, 1919–1942*. Cambridge: MIT Press, 1992. This collection of essays explores the impact of early-twentieth-century poster art on political, artistic, and cultural movements.

Web

http://www.coolmemes.com/reader/conroy.htm. Coolmemes's collection of Pat Conroy–related links includes audio clips, interviews, and biographical information.

http://www.crosswinds.net/~russ_posters/. This online gallery of Soviet, Czech, Polish, and Cuban communist propaganda posters offers good potential for comparison.

http://www.goarmy.com. The home page of the U.S. Army's recruitment Web site is shown in the text.

http://www.internationalposter.com. This is a commercial site selling war and propaganda posters, but it has a rather extensive catalog of both communist and democratic art available for viewing.

MASSACRE
Film and Fact

POEM, POSTER, ESSAY, 2 PHOTOGRAPHS, DRAWING, MAP

Tobe Hooper's film and Verlyn Klinkenborg's essay meet in Sherman Alexie's "The Texas Chainsaw Massacre," so you may want to save discussion of the poem until after you have provided historical context for it. Direct students to note how Alexie repeats the phrase "I have seen it and like it" in order to develop an ironic stance toward his subject matter: his attempt to "understand" Sand Creek through the lens of the horror film is actually meant as a condemnation of the act. Many individual lines in the poem have cryptic, thought-provoking undertones: "This vocabulary is genetic," "Violence has no metaphors; it does have reveille," and "I have been in love." You could also discuss in more general terms the attempt to use popular culture as a way of conveying the insights of one's personal culture, which is Alexie's admitted strategy here. Alexie sees himself as a true hybrid—note how the poem uses line breaks at the beginning and ending stanzas to literally show his dual identity as an "American / Indian."

Compare the tone of Alexie's poem with that of Klinkenborg's essay, which for the most part retains the objective distance of reportage style (using first-person pronouns sparsely). You may want to have students analyze the persuasive strategies at work in this essay, especially how Klinkenborg's apparent objectivity allows him to make subjective assessments (notice the disparaging adjectives he employs throughout). If the essay's final purpose is to raise awareness about Sand Creek and paint an unflattering portrait of Colonel John Chivington, does it do so effectively?

Again, in order to familiarize students with the intertextual references in the poem, you might consider showing excerpts from *The Texas Chainsaw Massacre*, keeping in mind that, even though nearly thirty years have passed, it remains one of the most gore-laden films of all time. In exploring the poster, ask students to point out elements that mark it as an ad for a horror film. Certainly, the threatening image is the most obvious element, especially when the saw's blade reaches past the frame of the picture and cuts into the title, as if to suggest its ability to reach the viewer. Also have students consider the textual elements of the poster, particularly the use of punctuation to imply excitement (the exclamation point) and suspense (the question mark).

CALLOUT QUESTION

Alexie may be referring to the fact that although the film is a work of fiction, it nonetheless provides representations of human suffering for purposes of entertainment. As Alexie is quoted in the comment in this section, he embraces his bicultural upbringing, valuing both his Native American heritage and the Western popular culture. Because of this viewpoint, he can easily see the profundity in looking at a lowbrow, B-grade horror movie as a metaphor for the very real tragedy at Sand Creek. For him, though the real event is without question more tragic than the fictitious film, both occur on the same continuum and serve as commentaries on our cultural affinity for violence.

MESSAGE

In his poem, Sherman Alexie draws a parallel between his enjoyment of the violence in the film and the bloodlust that drove the massacre at Sand Creek. His point is that our appetite for violence is cultural; by stressing that he likes the film, he shows the reader how easily one is persuaded to desire violence in any form. Ultimately, his tone could be read as ironic—his love for the film's violence does not overshadow the fact that Sand Creek was a real tragedy, and he subtly underscores that the actions at Sand Creek should be culturally condemned. At the end of the poem, he cautions against buying into the mind-set of accepting glamorized violence, this move from hunger to madness, because accepting violence even as fiction is not so far from committing violent acts.

METHOD

The poem's force doubtlessly depends on the reader's prior knowledge of at least the film, for one goal of the poem is to link a well-known text to a not-so-well-known event impacting Native American history. Knowledge of both the film and the Sand Creek massacre certainly add to the poem's impact, and after reading Klinkenborg's essay, readers may better understand why Alexie feels the need to remind us of this historical event from a Native American perspective.

MEDIUM

As a class project, students could work up storyboards for a documentary on the massacre at Sand Creek. You could break the class into different "production teams," each one responsible for a different task—sound, narrative, images, and so on. For preparation, have them view excerpts from similar documentaries such as Michael Apted's *Incident at Oglala* or Ken Burns's *Civil War*, and also provide them with examples of how storyboards are designed.

ADDITIONAL WRITING TOPICS

1. Are you a fan of the horror film genre? If so, list some of your favorite movies, explaining the qualities that you appreciate about them. If you don't care for horror flicks, explain which of their qualities you find uninteresting or objectionable. Then, compare horror films to a genre of film you do like.
2. Some of the statements in Verlyn Klinkenborg's essay perhaps suggest that he is either ambivalent about or does not like the way that Sand Creek has been memorialized. He describes the monument on the site as having "the head of a generic Indian in profile" and mentions how the site had previously been used as a dump and a place to smoke marijuana and drink. What sort of monument might Klinkenborg think more appropriate to this event? Can you find evidence in his text that supports your guess?
3. A quote by Sherman Alexie in this selection explains that the five primary influences in his life are his father, his grandmother, horror writer Stephen King, novelist John Steinbeck, and *The Brady Bunch*. Construct a similar list of your primary influences, including personal as well as popular sources, and provide rationales for the specific marks they have made on your identity.

DIVERGENCES

Print

Alexie, Sherman. *Indian Killer*. New York: Atlantic Monthly Press, 1996.

———. *The Lone Ranger and Tonto Fistfight in Heaven*. New York: Atlantic Monthly Press, 1993. This short story collection has been highly acclaimed.

———. *One Stick Song*. New York: Hanging Loose Press, 2000. This is Alexie's latest volume of poetry.

———. *The Toughest Indian in the World*. New York: Atlantic Monthly Press, 2000.

Custer, George Armstrong. *My Life on the Plains—Or, Personal Experiences with Indians*. 1874. Reprint, Norman: University of Oklahoma Press, 1977. The famous general's memoir may provide background for this selection.

Hilger, Michael. *From Savage to Nobleman: Images of Native Americans in Film*. Metuchen, N.J.: Scarecrow Press, 1995.

Web

http://www.sandcreek.org/. The Northern Cheyenne Sand Creek Massacre Site Project will provide background information for this selection.

http://www.texaschainsawmassacre.net. Set up by a dedicated fan, this exceptionally informative site about the film includes background on the story line, film locations, and actors, as well as a detailed FAQ.

Audio/Visual

Incident at Oglala: The Leonard Peltier Story. Directed by Michael Apted. Narrated by Robert Redford. 90 minutes, rated PG. Artisan, 1994. Videocassette. This documentary investigates the controversial murder of two FBI agents on the Oglala reservation and the equally controversial arrest and prosecution of Leonard Peltier for the crime.

Sherman Alexie. WHYY, Philadelphia, September 21, 1993. Audio recording. This interview with the writer was conducted by Terry Gross of National Public Radio's *Fresh Air*.

Smoke Signals. Directed by Chris Eyre. 89 minutes, rated PG-13. Mirimax, 1998. Videocassette. Alexie's screenplay tells the story of a journey by two young Native American men to retrieve their dead father for burial.

Texas Chainsaw Massacre. Directed by Tobe Hooper. 83 minutes, rated R. MPI Video, 1993. Videocassette. This 1974 cult classic is the original of the slasher film.

Forgetting Would Be a Second Abandonment
United States Holocaust Memorial Museum

ADVERTISEMENT

The German philosopher and aesthetic critic Theodor Adorno was noted for saying that to write poetry after Auschwitz is barbaric; his ideas on the creation of art in response to the Holocaust have been considered controversial. Similar controversy surrounded the construction of the United States Holocaust Memorial Museum. Many Holocaust survivors had misgivings about the way in which their ordeal would be remembered—they wanted the museum to do justice to their experience. Now, however, the USHMM is one of the most popular attractions in Washington, D.C., bringing in hundreds of thousands of visitors annually, and many people have applauded the museum's powerful efforts to memorialize the event. Some of your students may have visited the museum themselves, so you may want to ask them if they would be willing to share their experiences with the rest of the class. You could also spend time exploring the museum's Web site and discussing how effectively it defines the Holocaust for those who did not experience it. Finally, referring to the following selection on Maya Lin's *Vietnam Veterans Memorial*, you might ask students to discuss the respective roles of sculpture and the museum as memorials for awful events. How does each form attempt to create a fitting tribute to its subject matter?

You should eventually turn attention to the ad itself, analyzing how its structure mirrors the solemn tone projected in its message. Direct students to note how an air of seriousness is designed into every element of the ad, from the simple typography to the grim expression on the children's faces, from the judicious use of color to the stabilizing composition around the horizontal axis. You might finally point out how these features mirror the declarative language in the headline and the quote underneath the picture, not to mention the purely informational ad copy.

MESSAGE

The ad urges its American audience to remember the Holocaust of World War II as a moral plea, and it assumes that readers would be familiar with the United States' initial reluctance to enter into the war. Some people believe that this reluctance—the implied first "abandonment"—allowed Nazi Germany to commit large-scale genocide of European Jewry for many years. The ad explicitly asks us to remember the victims, and perhaps implicitly asks us to morally condemn the actions of the Nazis so that we are not in danger of repeating history. You might ask your class to come up with a list of various ways we can "remember" an event such as this.

METHOD

Although the ad carries a headline, there is no explanatory caption—the image, rather than any text, speaks to the reader. This particular image, of Jewish children adorned with the Nazi-mandated yellow Star of David, might have been chosen in order to emphasize the need to remember the Holocaust into the future—the commonplace equation Children = Future is implied. Also, the seemingly healthy children evoke sympathy in the reader. The ad's designers could have chosen one of many more atrocious images from the Holocaust—walking skeletons, mounds of bodies, dismemberments—but their purpose was to make us feel compassion for the survivors of this ordeal rather than to shock or horrify us.

MEDIUM

An inventive in-class exercise would be to have groups of students design their own ads, mixing images and copy text in order to produce the desired effect in the audience. Begin by having them analyze a few existing ads for the U.S. Holocaust Memorial Museum other than the one included in the text (see the museum's Web site). After establishing their own criteria of what makes an effective or ineffective ad, they can apply these standards to their own final product.

ADDITIONAL WRITING TOPICS

1. Freewrite about this image, including any other associations with the Holocaust that it brings to mind. This will allow you to get your own emotions on paper as well as help you examine our cultural memory about this event.

2. Why is it important that we remember tragedies and atrocities? Why not simply honor accomplishments and heroism instead? Write an essay that explores these questions.
3. Analyze the design of the U.S. Holocaust Memorial Museum's Web site (http://ushmm.org /museum). How are graphics, text, and audio combined, and what are the intended effects of these combinations? How would you describe the overall tone of the site, and how does it compare to the ad included in your reading? Choose a corporate site that advertises a product, such as that of Nike, McDonald's, or another large company. What similarities and differences are there between the for-profit Web site selling a product and the nonprofit Web site advertising the museum.

DIVERGENCES

Print

Adorno, Theodor. "Commitment." In *Art in Theory 1900–1990: An Anthology of Changing Ideas*, edited by Charles Harrison and Paul Wood. Cambridge: Blackwell, 1992. Adorno elaborates on his earlier claim that modern art after the Holocaust is incapable of elevating humanity as it had before.

Frank, Anne. *The Diary of Anne Frank: The Critical Edition*. New York: Doubleday, 1989. Critical commentary accompanies the widely read diary of Frank, a young Jewish girl who hid with friends and family in an attic during the Nazi occupation of the Netherlands.

Wiesel, Elie. *Night*. New York: Bantam Books, 1982. In this moving memoir, Wiesel recounts his own experiences with the Holocaust as a Jewish child.

Web

http://www.ushmm.org/museum. The United States Holocaust Memorial Museum Web site contains numerous online exhibitions in addition to information about the museum itself.

Audio/Visual

Schindler's List. Directed by Steven Spielberg. Starring Liam Neesom, Ralph Fiennes. 197 minutes, rated R. Universal, 1993. Videocassette. This is Spielberg's epic attempt to memorialize the true-life Holocaust story of German factory owner Oskar Schindler's attempts to save Jews from genocide.

MEMORIAL
Maya Lin

PROPOSAL, SKETCHES, ESSAY, 3 PHOTOGRAPHS

Playing off the definition of "memorial" included in this selection, start discussion by asking your class to participate in a free-association writing exercise where they list terms and ideas that come to mind when they encounter the word. Have them share their lists as a group and try to place their words into categories—for example, emotional, political, practical, personal, public, aesthetic. You may then extend the book's rather open-ended definition into one that accounts for the various cultural associations your students have listed.

For comparison, have your students research other memorials by going online or to the library, locating examples that commemorate either American wartime events or those from other nations. You might consider asking them how Lin's creation tries to de-politicize the event it memorializes, and how other monuments make political statements. You should also discuss one of the striking aspects of Lin's wall: the fact that it is nonrepresentational (i.e., that it does not try to look like anything from physical reality) is a significant departure from typical war monuments, which depict soldiers in the midst of battle or figures in otherwise tragic poses. This decision is doubtless a result of the complicated emotional climate surrounding the Vietnam War; if you have not already explored the public divisiveness during the war when dealing with other selections, it might be a good idea to do so here.

This selection gives the reader a kind of road map for how a piece of art comes into existence, a journey from its inception (the proposal) to its realization (the photographs of the actual wall) and even beyond to a point of distant reflection (the essay, nearly twenty years after the fact). Taking the various pieces of this selection into consideration, have your students debate the essence of an artistic creation. Does the initial proposal, the (almost) pure product of Lin's imagination, adequately capture the artistic reality of the wall? Note that Lin herself mentions surprise at how closely the final product mirrored her vision. Is the essence in the wall itself, its materiality? As the artist suggests, is it only by experiencing the wall that its intended aesthetic/social function is realized? If so, how do visual representations of this wall—the still photographs (especially the aerial shot, a perspective that Lin herself would likely frown upon); the digitized "walking tour" offered by the Web site; the countless depictions of the wall in melodramatic paintings, staged photos, and films—change how we as a society encounter the memorial?

CALLOUT QUESTION

Certainly, the feedback about the "naive" quality of Maya Lin's architectural sketches makes it likely that the verbal description played a large role in the jury's decision. The function of the description was to serve as a virtual substitute for experiencing the memorial—this function involves much more than the sense of sight, and you might want to explore with your students just how Lin's description sets out to imagine that experience. This question could lead into talk about whether or not Lin's description encouraged a specific way of "reading" the memorial, even though her original intentions were to construct as open-ended a memorial as possible.

MESSAGE

Maya Lin initially envisaged the *Vietnam Veterans Memorial* as a wound in the earth, as if some large knife cut a slit into the hillside. The result of this slice is the wall itself, which Lin wanted to seem less like massive blocks of granite and more like a symbolic portal, or "interface," into the memorializing process. Lin chooses "highly polished black granite" in order to render the surface "reflective and peaceful." Her description of the memorial focuses on the phenomenological, or experiential, encounter. She describes how one might interact with the wall (walking along its length, touching it); she describes the expected result of such interactions (the magnitude of names gives an impression of wholeness to the memorial); and she describes the work in its proper context rather than as an abstraction (in relation to the Washington and Lincoln memorials, and she mentions how the wall would act as a sound barrier for the nearby roadway).

METHOD

Lin's comparison of the memorial to a book is suggestive of a personalized experience—reading a book is a private, almost intimate activity, and Lin wants this experience to carry over into what is certainly a much more public text. She also likens the wall to an interface, perhaps in a nod to the digital medium, or perhaps in its unadulterated sense. Again, Lin's writing allows her to creatively conceive of the monument as an experience rather than as a static work of art. Drawing allows her to express only a visual rendering of the monument, but writing lets her describe the wall as it unfolds in space and time to a visitor. The power of writing in this instance is not lost on Lin, who felt that her description was a deciding factor in her design being selected by the planning committee.

MEDIUM

The offered definition of "memorial" focuses more on function than on shape or form, and this is key to the concept. Discuss with your class the effectiveness of Lin's design in achieving its goal—does her creation "preserve the memory of a person," as the definition suggests? Should a memorial be judged by other criteria as well? For example, would our society find a statue of, say, Richard Nixon riding atop a dragon to be a proper tribute, even if it would certainly preserve our memory of him?

ADDITIONAL WRITING TOPICS

1. Compare your aesthetic experience of the *Vietnam Veterans Memorial* with other images and texts associated with the Vietnam War. You can draw upon documentary photographs taken during the war, feature films set during the period, or war memoirs written by veterans (such as Tim O'Brien). Which do you find more powerful: graphic depictions of the war or Lin's minimalist memorial? Explain.
2. Acknowledging the different sources of controversy Lin outlines in her essay, how would you design a memorial to the veterans of Vietnam that takes these particular audience concerns into consideration? Provide both a detailed physical and experiential description of your design as well as a rationale for your aesthetic choices.
3. In her essay "Between Art and Architecture," Lin writes of her memorials, "I consider the monuments to be true hybrids, existing between art and architecture, they have a specific need or function, yet their function is purely symbolic." How do you interpret this quote with respect to her *Vietnam Veterans Memorial*? Draw upon her description of how she conceived and developed the monument to support your interpretation.

DIVERGENCES

Print

Lin, Maya. *Boundaries*. New York: Simon and Schuster, 2000.

Web

http://thewall-usa.com. The "Vietnam Veterans Memorial Wall Page" is maintained by veterans of the 4th Battalion, 9th Infantry Regiment.

http://www.loc.gov/exhibits/treasures/trm003p.html. This Library of Congress site details the history of the *Vietnam Veterans Memorial*'s development.

http://www.thevirtualwall.org. This site offers visitors a virtual tour of the *Vietnam Veterans Memorial*.

Audio/Visual

Maya Lin: A Strong, Clear Vision. Directed by Frieda Lee Mock. Not rated. Ocean Releasing, 1994. This documentary follows the acclaimed sculptor's career as a commemorative public artist and offers insights into how Lin attempts to preserve cultural memory without being overtly political in her sculptures.

5 Dividing Lines

When introducing this chapter, ask each student to compile a list of societal issues that are divided along strict, opposing ideological lines. What similarities and differences do you find among the lists? What do the variations tell you about your students? This chapter focuses on the issues of gender, class, ethnicity, activism, and politics. How often do these subjects come up in your students' lists? Be sure to identify subjects that are not included in this chapter and refer to them when you can—students may have a better understanding of the issues in the text when they can relate them to their own ideas.

A common theme among these selections is how easy it is to draw simplistic conclusions that divide people needlessly. Barbara Kruger's pieces all confront certain barriers to logic. Look at the *Help!* piece, with its seemingly unsolvable predicament. Kruger's point is to illustrate how difficult some decisions can be yet how easy it is for those uninvolved to form an opinion.

Vince Aletti's piece addresses these ambiguities as well, as he ponders what the distinctions between masculine and feminine really are. Essentially, you should alert your students to how each piece examines the ambiguity that goes along with a seemingly black/white issue—an understanding of such ambiguity makes for intelligent argument and skillful writing.

TALKING TO YOU
Barbara Kruger

BILLBOARD, MAGAZINE PAGE, BUS SHELTER

It has been suggested that the art of Barbara Kruger embraces opposition. Have your students list and discuss the various dichotomies these pieces exemplify: public and private, commercial and artistic, subjective and objective, powerful and powerless, visual and textual. Extend the list to include other common dualities, such as male and female, rich and poor, fat and skinny. Discuss how these pairs work to privilege one term over the other, and then examine how Kruger manipulates the various rankings in her art. In what ways, specifically, do these opposites come together in Kruger's works, and how does her style upset the distinctions between them?

If you are interested in a less abstract opener, you could also talk about Kruger's technique of addressing viewers directly. For instance, ask your students how they react to the billboard headline "Don't be a jerk." Do they automatically assume that the billboard is speaking to them? If so, do they begin reflecting on their own behavior, searching for signs that they might indeed be jerks? What exactly do they feel compelled to do for the young woman in the *Help!* bus shelter ad? How is that feeling compounded by elements such as the emphatic style of the headline, the use of an audience-directed question at the end of the text, and the model's direct gaze? And, in *Look at Me*, how does situating text against the closely cropped head shots of the models illuminate unspoken feelings that readers might have when looking at similar images in popular magazines such as *Elle*, *Cosmopolitan*, or *Vogue*?

MESSAGE

The messages sent by Barbara Kruger's work are not always straightforward; they require some work on the part of the viewer. For example, the text of *Untitled (Don't be a jerk)* is quite clear, but the message behind it is hard to discern. Why shouldn't I be a jerk? How am I a jerk? It is also hard to understand the photo of a large group of people with no accompanying text. When the text is seen along with the mass of people, the message becomes less ambiguous, and it is easier to see an individualist slant to the work. So it is also with the following two pieces—the text and image alone are parts of the puzzle, and the combination of the two provides for a message greater than the sum of the parts.

METHOD

Kruger's pieces use provocative phrases and confrontational language (e.g., second-person pronouns) that is usually directed toward the viewer. The red-and-white headline style immediately catches the eye, especially when set against a black-and-white background image—indeed, the graphical components are secondary features in many of Kruger's pieces. The choice in typeface—a bold, clean, sans serif font—serves to slow the viewer's eye as he or she reads over the text. The overall effect is one of assault, an in-your-face commentary that commands attention.

MEDIUM

In having your students debate the artistic value of Kruger's work, remind them of the strategies employed by postmodern art in general—collapsing distinctions between high and low art as well as between commercial art and the kind of art fit for the museum. Kruger's work, because of the unconventional forms and genres she uses, has consistently been dubbed ironic by art critics, and it is in large part that irony (besides the fact that her pieces do not directly "sell" anything) that makes her work worthy of serious artistic study.

ADDITIONAL WRITING TOPICS

1. In the comment included in this selection, Barbara Kruger says, "Talking heads and pronouns rule, in the best and worst sense of the word. I'm interested in how identities are constructed, how stereotypes are formed, how narratives sort of congeal and become history." In an analytical essay, define the various identities constructed in one of Kruger's pieces. How would you characterize the "speaker" of the text? How does that speaker seem to characterize the "reader" of the piece?
2. Kruger explains that the verbal technique of direct address—of using second-person pronouns, in other words—has always been an objective of her work. Write a short personal reaction to Kruger's work: How do the pieces make you feel as a viewer, and how do you imagine the artist intended for you to feel?
3. Think about the traditional public texts from which Kruger borrows: billboards, bus shelter advertising, and magazine advertising. Compose an essay in which you compare and contrast these forms of communication with Kruger's artwork. How does her work mirror these forms, and how does it depart from them?

DIVERGENCES

Print

Kruger, Barbara. *Thinking of You*. Cambridge: MIT Press, 1999.

Web

http://www.arts.monash.edu.au/visarts/globe/issue4/bkrutit.html. This page contains several images and details from a 1996 installation exhibit in Melbourne.

http://www.geocities.com/SoHo/Cafe/9747/kruger.html. This fan site features a rather good short personal essay on Kruger's art that could be used as a model text in class.

Audio/Visual

Barbara Kruger: Pictures and Words. 28 minutes. Facets Multimedia, 1996. This short documentary on Kruger's work showcases two of her SoHo shows. It can be obtained through http://www.buyindies.com.

FEMINISM
Gloria Steinem

MAGAZINE COVER, ESSAY

The memory of the Columbine High School tragedy is probably still fresh with your students, and you might want to use this as a springboard from which to begin looking at Gloria Steinem's article "Supremacy Crimes." To avoid any potential emotionally tense moments, start off class with a freewriting exercise rather than group discussion where students can write out their impressions of the Columbine shootings—where they were when they first heard about the incident, immediate reactions upon hearing it, any related conversations with friends and family members, and their own personal explanations for why such an event happened in the first place. After writing, students can then debate whether or not Steinem's position adequately explains the influences behind Columbine and similar episodes.

Steinem's article is a classic example of an argumentative essay, and you may consider teaching it as such, pointing out the different rhetorical tools it employs in order to prove its main point. The second paragraph provides a list of examples similar to Columbine that, taken together, prove to the reader that Steinem's topic needs immediate consideration. The fourth paragraph begins with the claim "We know that hate crimes, violent and otherwise, are overwhelmingly committed by white men who are apparently straight." Note the use of the pronoun "we," a subtle attempt to draw the reader to the author's position. Immediately afterward, Steinem provides several examples of straight, white, male serial killers as evidence supporting her point; additionally, in later paragraphs, she cites authorities on the subject such as Elliott Leyton.

Steinem is careful to acknowledge counterargument as well, which we see when she concedes, "Men of color and females are capable of serial and mass killing," but then refutes the claim on the basis that such occurrences are disproportionately rare. She then asks her audience to consider hypothetical situations that showcase the media's reluctance to make gender, race, and sexual orientation overt issues in tragedies like Columbine—an effective technique designed to make the audience complicit with her position because it requires no essential proof on her part, just the willingness of her audience to "go along" with her point.

Eventually we see an explicitly worded thesis statement (called a delayed thesis, which functions by already establishing proof before having to present the reader with a potentially challenging claim), followed immediately by a solution to the problem—namely, a call to raise our male children to value qualities that do not contribute to violent behavior, "empathy as well as hierarchy." Discussing these various techniques and analyzing how they work together to construct a cohesive argument could prove a valuable exercise in raising students' awareness of the rhetorical nature of their writing, while giving them a vocabulary to use when talking about the argumentative process in future exercises.

To further contextualize Steinem's piece, look at the July 1972 *Ms.* cover with your students or, better yet, see if you can locate a copy of the magazine from its early days. Explore the magazine's content to discover the kinds of issues that were deemed important at the beginning of the feminist movement, and discuss which issues have changed, which have disappeared, and which have remained constant. Ask your students to identify feminist victories as well as ongoing feminist battles, and also ask them how these political viewpoints affect other groups identified by their differences, such as ethnic minorities, the lower class, or children.

MESSAGE

In part, the choice of Wonder Woman for the cover of *Ms.* may be playful editorial commentary. The magazine might be ironically pointing out patriarchal attitudes toward the concept of the strong woman: an unrealistic, cartoonish fantasy. Nonetheless, Wonder Woman serves as an icon of female empowerment, one that women across the country would recognize more readily than they would Gloria Steinem or Simone de Beauvoir; in short, the editors probably thought the colorful and familiar image would entice a certain category of readers to pick up the new magazine. The magazine's popularity was likely fueled by the fact that there was no direct competition. *Ms.* filled a void, a silence, in the 1970s magazine market, and young women with a growing political consciousness were eager to pick it up.

METHOD

Steinem wants to find a different way of looking at a tragedy like Columbine, one in which we examine the deep, underlying causes of violence rather than assigning blame for a singular act. Her solution, therefore, depends on fundamental, long-term changes in how our society is structured. She argues that our cultural attitude toward masculinity—which privileges aggression, competition, and occasionally violent behavior—is a basic problem that needs to be overcome (as she suggests, by raising our boy children more like our girl children so that they learn to value empathy). To prove her point, she highlights the fact that ostensibly straight white males, the ones most susceptible to patriarchal pressures, are by and large the ones most likely to commit indiscriminant hate crimes against the "weaker" members of society. Her point is not to cast blame, for that would involve pointing out a particular agent, but rather to focus her audience's attention on these fundamental problems with society.

MEDIUM

Toward the end of the article, Steinem raises questions about how the news media might have spun the Columbine story differently if the offenders had been blacks, gays, or women. Her point is that the media do not make a point of the fact that the Columbine shooters were straight white males; the media silenced these identity markers and instead focused their arm-chair psychological explanations on typical societal evils like violence in video games, films, and the like. Steinem argues that we should dig deeper than that. Discuss with your class how news organizations might handle such charges and possibly consider having them compose mock-up editorial responses.

ADDITIONAL WRITING TOPICS

1. In a short reflective paper, compose a list of memories, ideas, concepts, and anything else you can think of that you associate with the term "feminism." Follow up this list with an explanation of how these associations inform your current opinion of feminism.
2. Research several past issues from *Ms.* magazine's archives. How has the editorial content of the magazine changed over the years? Can you identify any broad cultural shifts that might account for these changes? Also, what topics remain constant for the magazine over time?
3. Gloria Steinem proposes several hypothetical incidents that the media would have treated differently from Columbine. She considers the situation from the position of race, sexual orientation, and gender. How might the news media cover these different scenarios? Choose one and write a mock news analysis that attempts to explain its causes.

DIVERGENCES

Print

Doty, William. *Myths of Masculinity*. New York: Crossroad, 1993. This volume explores the prevalent myths of Western culture from ancient Greece to contemporary times that help inform our conceptions of masculinity.

Steinem, Gloria. *Outrageous Acts and Everyday Rebellions*. New York: Holt, Rinehart, and Winston, 1982.

———. *Revolution from Within: A Book of Self-Esteem*. Boston: Little, Brown and Co., 1992.

Web

http://www.msmagazine.com/. *Ms.* magazine's Web site includes an abundance of archived material.

Audio/Visual

Boys Don't Cry. Directed by Kimberly Pierce. Starring Hilary Swank, Chloe Sevigny. 116 minutes, rated R. Twentieth Century Fox, 1999. Videocassette, DVD. This award-winning film is based on the true story of Teena Brandon, a transgender teen who was murdered after acquaintances discovered she was biologically female.

No Safe Place: Violence Against Women. Produced by KUED, Salt Lake City, 1998. Videocassette. This PBS documentary series includes an interview with Gloria Steinem. To order, call KUED Video-Finders at 1-800-343-4727.

GENDER
Male/Female

ESSAY, 2 PHOTOGRAPHS

Jesse DeMartino's photographs of Houston, Texas, skateboarders offer a perfect illustration of how we as a society structure ourselves in terms of gender. Though skateboarding has appealed to some women, it is by and large a culture of young men. Ask your students to list the qualities and attributes they associate with the figures in DeMartino's photos (e.g., aggression, bravado, physicality, competitiveness) and discuss how these terms help foster our understanding of masculine and feminine categories generally. Have them brainstorm a list of other subcultures that similarly define particular gender roles: the "gangsta" lifestyle, the gothic scene, or the neo-hippie

Vince Aletti's essay "Male/Female" argues that our sense of masculinity and femininity is culturally determined, and that this sense is in a constant state of agitation—new, oftentimes conflicting definitions of "male" and "female" show up in popular culture all the time. As a culture, we are nevertheless able to make sense of all these pieces, despite our concerns as individuals about how we might measure up to the benchmarks. Aletti proves his point by assembling a catalog of images, which he cynically refers to as "icons-R-us." He closes his final paragraph with a rhetorical question—"Who do we think we are?"—which he proceeds to answer himself—"A work in progress." This last line invites the reader to consider additions to Aletti's catalog, and this invitation could be fashioned into a variety of in-class exercises in which students list more (and more familiar) examples of images from our shared culture that train us how to "perform" a certain gender. Challenge your students to think of current celebrities, sports stars, films, musicians, or trends in fashion and slang that add to our ever-changing concept of gender.

Finally, discuss the concept of gender itself with your students. Have your students consider that perhaps gender is an essential characteristic of a person rather than a culturally constructed one. This position seems to be opposite the one Aletti offers, so you may want to compare the two opinions. Is Aletti's assertion that culture decides the gender of a child a solid position? Or is gender a quality that is somehow more pure and powerful than even biological sex? Or is it that, by virtue of the fact that gender is culturally encoded, we have the freedom to play with our genders and be whomever we want? You could even have your students respond to this by writing an essay beginning with the line, "To me gender is . . ."

MESSAGE

The pieces in this selection make conflicting claims about gender, but on the whole they work together to illustrate the extremely subjective and multiple ways that we all construct gender—there is ultimately no unified nature to gender, but rather a conglomeration of definitions to which we all contribute. While Aletti's view expresses the importance of social conditioning and contextualization in shaping gender roles. Jesse DeMartino's pictures show representations of gender, in this case adolescent males, that are performances without any explicit acknowledgment of the social script. The young skaters depicted here are busy being boys, and it apparently comes naturally to them.

METHOD

You may want to informally quiz your students on the various allusions Aletti makes in his essay, noticing which ones are familiar to them and which ones are not. You could also discuss those people and images from pop culture that challenge our traditional notions of masculine and feminine, from Dennis Rodman and RuPaul to Michael Jackson and Marilyn Manson.

MEDIUM

Aletti mentions Calvin Klein's influence on shaping gender norms; images of Kate Moss and Marky Mark are familiar from CK ads. Generally, fashion, sportswear, and cosmetic industries have a stranglehold on the gender-shaping market, and you might consider conducting an exploratory exercise in which your students sift through various magazines for ads that seem to define "male" and "female": Revlon, Clinique, Clairol, Tommy Hilfiger, Ralph Lauren, and Maybelline are just a few of the dominant advertisers. Part of the reason that these ads are so effective at conveying messages about gender is that they naturalize gender in order to sell their product. Because the motivation of an advertiser is different from that of an artist (advertisers want to be our friends more readily than artists do because they want to sell a product or image to us easily), ads are generally better than artworks at catching our attention—they appeal to rather than challenge our common sensibilities about gender.

ADDITIONAL WRITING TOPICS

1. Individually, take five minutes to jot down images and ideas that are traditionally associated with femininity and masculinity. Then, compare your notes with those of two or three other students. How are the notes similar, and how are they different? Why do you think this is?
2. Aletti asks, "Even if we accept the notion that, ideally, we share both masculine and feminine traits, what exactly are they?" What would you describe as your masculine traits, and why do you consider these traits masculine? Your feminine traits? How do you think you developed them?
3. The word "gender" is often used but seldom defined. How would you define it? Using your definition, reexamine stereotypical examples of male and female in the media.

DIVERGENCES

Print

Aletti, Vince. *Malefemale*. New York: Aperture, 1999. Aletti edited this collection of portraits, snapshots, collages, and other images exploring gender issues.

Web

http://www.aperture.org/. Jesse DeMartino's photos originally appeared in *Aperture* magazine.

http://www.nerve.com. This popular site offers untraditional and unflinching views on sex, gender, and relationships.

Audio/Visual

Gleaming the Cube. Directed by Graeme Clifford. Starring Christian Slater, Tony Hawk. 109 minutes, rated PG-13. 1989. Videocassette, DVD. This teen film starring a young Christian Slater and skateboard hero Tony Hawk has been prized because of how "cheesy" it is.

The Human Sexes: A Natural History of Man and Woman. Directed by Desmond Morris. 6 segments, not rated. Discovery Communication, 1997. Videocassette.

Paris Is Burning. Directed by Jennie Livingston. 78 minutes, rated R. Fox/Lorber, 1991. Videocassette. This acclaimed documentary focuses on the transvestite and drag-queen lifestyle in the inner city.

A Journey to Class Consciousness
bell hooks

ESSAY

Throughout her description of how her personal sense of class consciousness developed, bell hooks uses a journey metaphor as well as spatial language: "domain," "starting point," "where we stand." Construct an in-class exercise where you have your students map out (either visually or in writing) hooks's journey. Direct them to make note of the specific, physical locations in her journey—her mother's kitchen, the library, the university campus—as well as any lessons she might have learned along the way. Ultimately, students should be able to articulate hooks's thesis—that class is an identity marker that intimately affects, and is itself affected by, gender and race.

Social class is often a challenging topic to discuss because, as studies have shown, most Americans tend to self-identify as middle class, regardless of whether or not that is the case. Hooks, in fact, thinks that we feel a need to ignore class issues, and that we do so by obscuring them with other issues: "Race and gender can be used as screens to deflect attention away from the harsh realities class politics exposes." Have students examine hooks's notion that Americans are reluctant to make class a part of the discussion about social inequality. Does this position conflict with our idea of America as a land in which anyone, regardless of how poor, can become successful through hard work and opportunity, a land in which there is no established ruling class?

Discuss with your students the final paragraph of the essay, in which hooks predicts two possible outcomes for American society. In one, she sees an ever-escalating class divide that ultimately erupts into violent class warfare. The preferable alternative is a "class-free" nation with a "just" economic system. Have your students imagine in detail what these two opposing outcomes might look like.

MESSAGE

A class-conscious person understands his or her social and economic standing with respect to other groups in society—it is not simply a matter of thinking that you are part of the middle class, for instance. Class consciousness means being aware of the ways in which a person's socioeconomic status affords certain benefits or precludes certain opportunities. The term is borrowed from Marxist theory, which predicted the eventual revolt against the bourgeois state by the working class once the proletariats came to understand their class position. Have your students discuss how race and gender are intimately related to class and how being part of a privileged group generally results in privileged class status and vice versa.

METHOD

Though by now hooks is part of the upper class, in terms of both her professional and her economic standing, her writing still reflects the thinking that emerged during the awakening of her class consciousness as a lower-class member of society. Still, she would argue, two facets of her identity—being African American and being a woman—will continue to keep her class standing somewhat marginalized. As a result, she goes on exploring the question of how class is influenced by factors such as gender and ethnicity.

MEDIUM

As hooks suggests, the news media tend to hide class markers when covering events that obviously have class issues at their center. In fact, the choice of whether to cover an event at all is often dependent on class (she wonders if O. J. Simpson's case would have been the spectacle it was if it did not involve a wealthy celebrity). Hooks condemns as socially irresponsible the media focus on factors like race alone because such a focus creates somewhat manufactured social tensions and glosses over the underlying class differences that create conflicting racial identities.

ADDITIONAL WRITING TOPICS

1. In what social class would you place yourself? In what ways does this classification shape the way you view society? Be sure to cite specific examples to help explain your points.
2. In her second paragraph, hooks cautions her audience not to look at the O. J. Simpson trial as if it were "all about race" and even goes on to suggest that "if O. J. Simpson had been poor or even lower-middle class, there would have been no media attention." Describe a similar event that evolved into a media spectacle, and explain how race, class, and gender were interrelated factors in its formation.
3. List the various ways—beyond race, class, and gender—in which you can identify yourself. Come up with terms as specific or as general as you like. Which labels are the most important in your own life, and why?

DIVERGENCES

Print

Breslin, Jimmy, and Stanley Crouch. "The Bad News: The Good News." *Esquire* (December 1995): 108ff. Offers two point/counterpoint opinions on the O. J. Simpson trial.

hooks, bell. *About Love: New Visions*. New York: William Morrow, 2000.

———. *Ain't I a Woman?: Black Women and Feminism*. Boston: South End Press, 1981.

———. *Bone Black: Memories of Girlhood*. New York: Henry Holt, 1996.

———. *Feminist Theory: From Margin to Center*. Boston: South End Press, 1984.

———. *Killing Rage: Ending Racism*. New York: Henry Holt, 1995.

———. *Outlaw Culture: Resisting Representations*. New York: Routledge, 1994.

———. *Talking Back: Thinking Feminist, Thinking Black*. Boston: South End Press, 1988.

———. *Where We Stand: Class Matters*. New York: Routledge, 2000.

———. *Wounds of Passion: A Writing Life*. New York: Henry Holt, 1997.

———. *Yearning: Race, Gender, and Cultural Politics*. Boston: South End Press, 1990.

Web

http://voices.cla.umn.edu/authors/bellhooks.html. The University of Minnesota's "Voices from the Gap" site includes a relatively complete online fact sheet for hooks, along with links to other pages, a bibliography, biographical data, and critical commentary.

Audio/Visual

bell hooks on Video: Cultural Criticism & Transformation. 70 minutes, not rated. Distributed by Facets Multimedia through http://www.buyindies.com. In this two-part documentary, bell hooks argues for the transformative power of cultural criticism.

California vs. O. J. Simpson. 2 volumes (465 minutes), not rated. MPI Home Videos, 1995. Videocassette.

Christopher Darden. WHYY, Philadelphia, March 31, 1996. Audio recording. On this episode of National Public Radio's *Fresh Air*, host Terry Gross interviews Darden, a member of the O. J. Simpson prosecutorial team.

TURF WAR
Gangs

ESSAY, 4 PHOTOGRAPHS

This selection continues the discussion begun in bell hooks's essay on how the socioeconomic class of a particular group of people helps determine their depiction in the news and entertainment media. Ask your students to note some of the stereotypical characteristics of an urban gang member. Also ask them what sorts of texts helped construct this image for them: specific films, news stories, music albums, or videos, for instance. You could follow this discussion by turning first to Joseph Rodriguez's pictures—what about this subculture is Joseph Rodriguez able to say through his photography? Is it simply all violence for the sake of violence, or is there a humanizing aspect to the lifestyle as well? How do these pictures uphold the gang member stereotype, and how do they complicate it? Have them take note of such elements as the posturing machismo of the Florencia 13 gang members as well as the more disturbingly sentimental depictions of family life in the picture of Mike Estrada, who longingly holds a picture of his jailed father, and of Chivo, who teaches his daughter how to use a handgun. (Also, take this opportunity to compare this family scene to one of Norman Rockwell's.)

You could ask the same questions of Richard Rodriguez's essay. It should be readily apparent that the author at least ostensibly upholds the stereotype—in the first paragraph alone, we are bombarded with a catalog of gang culture accessories, delivered in a disgusted tone: "Oh, how I hate their stupid sign language, occult and crooked palmings, finger-Chinese." It should be noted, however, that this is done for rhetorical effect, that Rodriguez's main point is to portray himself as one of "us," non–gang members, by mimicking what he imagines are the language and thoughts of this group. The essay moves from a position of pretend moral indignation for the gang life to one of genuine concern for mainstream society's inability to empathize with or want to help this group escape the dual cycles of violence and poverty. Eventually, Rodriguez makes his words ours—we become the ones yelling "MONSTERS! ANIMALS!" while at the same time devoting high-class fashion shows to "gangsta chic" styles, pumping iron in exclusive gyms while listening to NWA, and celebrating the drive-by as it is splashed up on the silver screen. In the end, Rodriguez faults those of us in mainstream society for believing too much in "the first-person singular pronoun," for not believing in souls, for being too much involved in our "pecs and abs" to reach into the inner cities from the safety of our suburbs and pull out the children who live in danger every day.

As a model multivocal essay, "Gangstas" effectively combines street talk with conventional essay prose, seamlessly moving from such statements as "What I like best about Joe Rodriguez's photographs is that they are devoid of middle-class *nostalgie de la boue*" to "It would be a smart idea not to look at them—no, I mean it—these are children, but they are children with machetes and guns and no point of ref-france, so betta show me deffrance, or you are outta bref, once I blow your brains to hell." You could use Rodriguez's essay as a model for a writing activity encouraging students to use more than one register of language within a single essay.

MESSAGE

Richard Rodriguez finds the gang culture to be an anomaly in our society, which values individuality above all else; as he puts it, "We are people who believe in the first-person singular pronoun." Yet the inner-city gang does not provide an especially healthy model of a communal lifestyle because of its violent propensities. Rodriguez would prefer to see a shift to communal ways of being in which we would be able to comprehend "our lives in common."

METHOD

Certainly, Joseph Rodriguez's photographs do not romanticize their settings—there is a disturbing quality to the image of the father handing his infant daughter a handgun as a smiling mother looks on, and the photo of the dead child in the coffin is nothing short of tragic. At some point, the question of whether it is Richard or Joseph Rodriguez who portrays gangs better comes down to the question of medium, but while the essay uses the stereotypes to make the point that mainstream culture does not completely understand gangs, the photographs might be read as moving *beyond* the stereotypes of gang life.

MEDIUM

Richard Rodriguez admits to hating an entire litany of gang culture aspects: hand signals, dress code, hairstyles, tattoos, jargon, and especially rap music. Actually, he uses the rap music genre as a metaphor for the philosophy of the gang lifestyle, calling it the "rhymed-dictate" that "encourages rhythmic slogan-ing and jingle—monotony posing as song, attitude posing as thought." While Richard Rodriguez points out mainstream society's hypocrisy toward urban gang life (we glamorize it while vilifying it), he does not necessarily take that extra step of showing the reader how gang culture is different from the distanced stereotypes we have created—that is the work of Joseph Rodriguez's photographs.

ADDITIONAL WRITING TOPICS

1. In the final paragraph of his essay, Richard Rodriguez draws a connection between the gang culture, which disturbs him, and his own gym culture. How does the author make this parallel, and what other similarities between these two worlds can you find throughout the essay? What other subcultures bear similarities to gang culture?
2. Mexican Americans and Hispanic Americans are recognized as the fastest-growing minority groups in the United States and "Hispanic" is soon expected to become the largest ethnicity over the next few decades. What signs of Hispanic influence in America do you notice today? Describe as many of these influences as you can.
3. Write a short but detailed narrative based on one or more of Joseph Rodriguez's photos included in this selection. Attempt to capture in your writing the tone you think Rodriguez intended in his photography.

DIVERGENCES

Print

Rodriguez, Joseph. *East Side Stories: Gang Life in East L.A.* New York: Powerhouse Books, 1996.

———. *Spanish Harlem*. Washington, D.C.: National Museum of American Art, 1994.

Rodriguez, Richard. *Days of Obligation: An Argument with My Mexican Father*. New York: Viking, 1992.

———. *The Hunger of Memory: The Education of Richard Rodriguez*. Boston: D. R. Godine, 1982.

Web

http://eclipse.barnard.columbia.edu/~as833/bc3401/. This hypertextual essay by a Columbia University student on gang culture makes for an excellent model of online writing.

http://www.gangsorus.com/gangs/ganggirls.html. The "Gangs or Us" Web site has a lot of interesting anti-gang content (the kind that might run counter to Richard Rodriguez's sentiments), but this page in particular deals with the topic of female gang members and includes several related links.

Audio/Visual

Boyz N the Hood. Directed by John Singleton. Starring Ice Cube, Lawrence Fishburne, Cuba Gooding, Jr. 112 minutes, rated R. Columbia/Tristar, 1992. Videocassette. Three childhood friends grow up in South Central Los Angeles and experience gang violence.

Menace II Society. Directed by Allen and Albert Hughes. Starring Larenz Tate, Tyrin Turner. 104 minutes, rated R. New Line Studios, 1993. Videocassette, DVD. This film offers a violent (and some say frank) look at the gang scene in Los Angeles.

Wanderers. Directed by Philip Kaufman. Starring Ken Wahl. Rated R. Warner, 1992. Videocassette. Originally released in 1979, this film is set in the Bronx in 1963.

West Side Story. Directed by Robert Wise and Jerome Robbins. Starring Natalie Wood. 151 minutes, not rated. MGM/UA, 1961. Videocassette, DVD. This famous musical is an adaptation of the Broadway play, in turn a version of Shakespeare's *Romeo and Juliet*. Set in New York City and pitting Sharks against Jets, the story provides background for a comparison of gang life then and now.

ACTIVISM
Groups with a Cause

3 ADVERTISEMENTS, LETTER

One way you might begin thinking about the three advocacy advertisements in this selection is by comparing them in terms of composition. Considering that all three have similar goals—to generate public awareness about social issues they find important—their compositional approaches are somewhat different. The most obvious difference is the ratio of text to graphics in each ad. While the American Indian College Fund and People for the Ethical Treatment of Animals advertisements have very little copy text to accompany their striking visual elements, the National Campaign Against Youth Violence ad has text at the very top and uses striking photos of domestic violence in place of a child's eyes. The AICF's image is expansive, and the open sky comprises over two-thirds of the entire image—perhaps this choice of cropping (which some might criticize for having too much "dead space") is meant to be suggestive of the wide-open potential of the Native American population, which is embodied by the figure situated in the bottom right-hand corner. Have students focus on this figure, which is partially blocked by shadow and is posed in a way open to interpretation—is the figure blissful, anguished, triumphant, messianic? Likewise, the PETA ad's visual elements mirror its message. The horizontal composition defined by the fence rails is repeated in the caption's layout against a block of color; the overall effect is to emphasize the idea behind the bold word "victim" in the caption: captivity, exploitation, murder.

The balance between the verbal and the visual shifts in the NCAYV ad. The copy text in this ad relies on irony to convey its message—"Children aren't born violent, but you can certainly change that"; the copy is situated directly above several scenes of violent parental abuse, the irony being that parents can change violent behavior for better or for worse. These images run across the middle of a young boy's face, documenting the reality of what many children witness every day. This choice of image placement is intentional, as we see images compounding one another, thus increasing the immediacy of the message. You could also examine the ad's rhetorical efforts to establish a sense of authority in its verbal component: by using simple, declarative sentence structures directed at "you," it implicates "us," all adults. The ad ends with "Is there any way to stop youth violence? Try starting with yourself." The ad is a direct call to all parents, all adults, rather than an abstract statement of a problem.

Finally, Timothy McVeigh's letter can be used as an example of a position statement that does not frame the issue of veganism in simplistic "either–or" terms, but rather discusses the killing of animals for food in terms of degree—though McVeigh does not care for the practice of mass-market consumption of meat, he does "believe in the *reasonable* taking and eating of game." He then goes on to challenge the logic of the vegan lifestyle (a radical version of vegetarianism in which no animal products of any sort are consumed or used) by suggesting that the lines vegans draw are ultimately arbitrary ones. Though McVeigh's argumentative technique is a bit sophisticated, it is not without its rhetorical problems. It is nearly impossible to ignore who is speaking in this case; this letter provides an excellent opportunity to discuss with your students the persuasive capacity of a speaker or writer's ethos. Is the audience likely to buy into McVeigh's argument, regardless of its content, considering the fact that he was (and remains after his execution) one of the most hated figures in America?

MESSAGE

In keeping with PETA's aim to get people to stop using animal-based products, this ad tries to dissuade viewers from buying leather goods. Without being graphic, the ad (with its cute calf) reminds its audience that the manufacturing of leather requires the killing of animals. Most of us know that leather is the skin of an animal, but given the prominence of leather goods in the marketplace we tend not to think about how they are made.

METHOD

The question in the AICF ad is rhetorical, the implication being that the reader has not in fact seen a "real" Indian. The purpose of the AICF campaign is to raise awareness of indigenous peoples in a society that caricatures tribal warriors as baseball team mascots. Beyond raising awareness, another of the AICF's aims is to provide Native Americans with financial assistance for college. The idea of this ad is that by helping the AICF, you (a non–Native American) will be gaining an opportunity to interact with "real Indians" because of their successful foray into your world.

MEDIUM

Given the amount of product ads we see in our lifetime, it comes as no surprise that advocacy ads would take a familiar approach so that we can readily distinguish them from the other contents of a magazine or newspaper. The AICF in particular is successful in styling itself after a product ad, for the image of a man with his arms beatifically stretched out recalls so many ads that claim their product will elicit a similar response, thus initiating "branding"—if you associate with their product, you too will feel terrific. The PETA ad is also successful in mocking product ads, with its heavy emphasis on image and simple text.

ADDITIONAL WRITING TOPICS

1. Research other print and television advocacy advertisements that you find effective and explain what makes them work. Consider whether it is the message itself that is compelling to you, or whether it is the way the message is delivered.
2. Two of the texts included in this selection—the letter from Timothy McVeigh and the Youth Violence ad—make emotional appeals to the

audience with regard to the people represented in them. In a short analysis, explain how the inclusion of these people functions rhetorically.

3. Might any of the advertisements in this selection persuade you to think or act differently? If so, indicate how. Is this the intended effect of the ad? In a group, come up with an important issue and construct an advertisement in the style shown here. After analyzing the PETA and NCAYV ads, use their strategies for your own purposes.

DIVERGENCES

Print

Garcia, Leonardo. *Advocacy Advertising*. Manila: De La Salle University Press, 1999. Written for students of advertising, this textbook explains the methods involved in analyzing audiences (public relations initiatives, focus groups, etc.) specifically for developing socially conscious advertising.

Web

http://www.collegefund.org/main.shtm. The American Indian College Fund's site, too, has a definite style.

http://www.noviolence.net/. The National Campaign Against Youth Violence's Web site has its own distinctive style.

http://www.peta.org/. PETA's Web site contains many images in the style of the one shown in this selection.

The Cult of Ethnicity, Good and Bad
Arthur M. Schlesinger Jr.

ESSAY

Arthur Schlesinger believes that cultural assimilation is in the best interest of the United States. Further, he believes that the nation's unique ability to include various racial and ethnic groups is a strength that the rest of the world, especially those countries experiencing religious or racial strife, should learn to emulate. In making this argument for the melting-pot vision of America, Schlesinger presents our current cultural preoccupation with multiculturalism and political correctness as a passing fancy, as "hullabaloo," and yearns for an educational curriculum that does not portray Western European civilization as the "root of all evil." After having your class distill the basic thesis of Schlesinger's argument, have them discuss its viability. Ask them to consider the importance of developing a curriculum to include the cultures of other racial, ethnic, and religious groups. Does the concept of "American" in any way benefit from the different perspectives and worldviews offered by such an education?

Schlesinger's piece carries with it the potential for volatile debate, so to avoid divisiveness, you might want to have your class focus on how the essay's argument is structured. Have students map out the author's various persuasive techniques, such as establishing the worldwide scope and overall importance of the topic, citing numerous negative examples of multiethnic nations, redefining multiethnicity by drawing on historic (and by default authoritative) quotations by John Adams and Michel-Guillaume-Jean de Crèvecoeur, and making the not-so-subtle association between an Afrocentric curriculum and the Ku Klux Klan. You could also discuss with your students the dangers inherent in slippery slope reasoning—Schlesinger's assumption that trends in multiculturalism inevitably will lead into radical separatism on par with that in Israel or the former Yugoslavia might be seen by some as an alarmist argument.

MESSAGE

The word "cult" definitely carries negative connotations when used by Schlesinger, and the implication is that the current cultural fascination with different ethnic, racial, and spiritual backgrounds is being forwarded by a vocal but nonetheless small minority. He writes: "Most American-born members of minority groups, white or nonwhite, see themselves primarily as Americans rather than primarily as members of one or another ethnic group." Consequently, he sees this "cult of ethnicity" being developed by overzealous educators and legislators with unacknowledged separatist tendencies.

METHOD

Schlesinger sees the spirit of America as ever-changing, and that change is brought about by the inclusion of new cultures—this has been going on since the country's inception. Ultimately, the goal of American culture, as he would like to see it, is assimilation.

MEDIUM

Schlesinger's tone lends a measured, authoritative air to his argument, which somewhat downplays the fact that his position is a politically conservative one. Any group with an interest in preserving the cultures or forwarding the civil rights of certain minority groups might find this essay a bit offensive—such groups may include the NAACP, NOW, or the ACLU.

ADDITIONAL WRITING TOPICS

1. A familiar, long-held metaphor of the United States' multiethnic composition has been the melting pot, though some sociologists and multicultural scholars have argued for a different metaphor that highlights the cultural differences that remain throughout our culture—the tossed salad or the patchwork quilt, for instance. Which of these metaphors (or another you can think of) represents the reality of American culture? Which one do you find more preferable? Write an essay that explains your answers to these questions.
2. The current motto for the United States is *e pluribus unum*—"out of many, one." Imagine that you were part of a task force charged with developing a new motto or slogan. What would you choose? Be sure to include a detailed rationale for your decision.
3. Based on evidence gathered from Schlesinger's argument as well as your own observations, what are some of the lessons that other countries can learn from the United States in order to develop more amiable relations between their different ethnic or religious factions? Propose several solutions.

DIVERGENCES

Print

Franken, Al. *Rush Limbaugh Is a Big Fat Idiot and Other Conversations*. New York: Delacorte Press, 1996. This book contains the *Saturday Night Live* humorist's observations on conservative politics.

Limbaugh, Rush. *The Way Things Ought to Be*. New York: Pocket Books, 1992. This best-seller represents the ultraconservative pundit's take on various issues.

Schlesinger, Arthur. *The Age of Jackson*. New York: Book Find Club, 1945.

———. *The Disuniting of America*. New York: Norton, 1992.

———. *A Thousand Days: John F. Kennedy in the White House*. Boston: Houghton Mifflin, 1965.

Web

http://www.washingtonpost.com/wp-srv/national/longterm/meltingpot/melt0222.htm. This URL leads to the *Washington Post*'s series of articles on the "Myth of the Melting Pot," which discusses the idea of cultural assimilation from a variety of ethnic and socioeconomic positions.

JUST DO IT
www.whatsyourantidrug.com

2 ADVERTISEMENTS, 3 SCREEN SHOTS

By now, you may have encountered several selections in *Convergences* where slogans and mottoes play a key role. Begin discussing this selection by looking back over some of these slogans—*e pluribus unum*, "Only her hairdresser knows for sure," "Be all that you can be"—and then compare them to "What's your anti-drug?" How do these different phrases measure up against one another in terms of initial impact, memorability, and longevity? Extend this exercise into a brainstorming session in which the class randomly lists the many slogans, mottoes, and catchphrases that saturate our popular culture. Follow this list up with a discussion of why we find these sayings so effective. What are the criteria for a good slogan? Perhaps even have your students try a hand at composing their own slogans for mock campaigns that you assign to them.

When looking at the National Youth Anti-Drug Media Campaign's ads, ask students to discuss the significance of the mosaic-style graphics—meant in part to suggest to the viewer that the audience is a wide cross-section of young people with varying hobbies and tastes. In effect, the faces in the ads become composites of an entire youth demographic. How are these groups identified, and what do the tiles featuring icons ("food," "video games," "sci-fi") say about the makeup of these different groups? Finally, what type of ethos is conveyed in the text of the second ad, with its heavy use of fragments and imperative sentences? Rather than an assertive, directive tone similar to those you would expect from ads trying to sell consumer products, it promotes a do-it-yourself mind-set that stands apart from the herd mentality fostered by traditional advertising strategies. You might point out how the handwritten script also helps foster the "by youths, for youths" image of the campaign.

Spend some time exploring the "What's Your Anti-Drug?" Web site, and maybe even have your students jot down notes documenting their experiences in using the site. They could focus on which design features of the site appeal to them at first glance, what navigational choices they make, where they get lost/bored/excited, and so on. Afterward, ask them to compare the Web site and the print ads, with a focus on how each medium affects the message. For instance, the interactive Web site allows the viewer to vote for favorite anti-drugs or submit personalized content, acts that the static print ads encourage but are incapable of facilitating.

MESSAGE

One consistent observation should be the difference in the levels or registers of language used in the kids' postings (more casual) and the content provided by the National Youth Anti-Drug Media Campaign. The campaign organizers include content by children to create a sense of commonality and community between the user and those kids who have already contributed to the site—thus making the site's message much more appealing.

METHOD

Students might note the honest, straightforward tone used in the writing on the Web site, an "in-the-know" type of ethos that tries not to be cheesy. This tone is supported by the hip visual elements, which employ a sophisticated sense of design. For example, the colors are not overpowering primary hues, the site's use of rollover graphics is not overdone, and the graphics are not childish illustrations. In short, the entire campaign attempts to treat its audience as more than mere children.

MEDIUM

The success of the Web campaign depends on interactivity. With high-risk teens, however, a more unidirectional medium such as television or radio (what some media theorists refer to as "push" technologies) might be more effective, because this group might not be as receptive to actively seeking out the Web site, not to mention buying into its message.

ADDITIONAL WRITING TOPICS

1. In a short expository essay, write your own answer to the question posed by the "What's Your Anti-Drug?" campaign. Describe the hobby, activity, or item that represents your "anti-drug" in as much detail as possible, and convey to the reader why it is that you enjoy it.
2. Compare and contrast this ad campaign with one that has different goals, such as a political candidate's election campaign or a commercially oriented campaign trying to sell a product. Analyze how these different goals affect the various components—slogans, graphics, logos, and overall message—of each campaign.
3. In a group of three or four, design an anti-drug campaign targeted at high-risk groups. Decide which media you will use and where you will advertise.

DIVERGENCES

Print

Gerber, Jurg, and Eric L. Jensen, eds. *Drug War, American Style: The Internationalization of Failed Policy and Its Alternatives*. New York: Garland, 2001. This recent collection of essays offers various globally oriented opinions on how the war on drugs ought to (or ought not to) be fought.

Web

http://www.thetruth.com. The Truth is an organization aimed at reducing underage smoking. You might also seek out some of this organization's striking ad parody commercials from the past two years.

http://www.whatsyourantidrug.com. The Freevibe Web site is sponsored by the National Youth Anti-Drug Media Campaign.

Audio/Visual

Traffic. Directed by Stephen Soderbergh. Starring Catherine Zeta Jones, Michael Douglas, Benicio del Toro. 147 minutes, rated R. USA Films, 2001. DVD. This Academy Award–winning film (set in Ohio, California, and Mexico) addresses the complicated political, social, and personal aspects of the war on drugs.

ELECTION 2000
Bush and Gore

ESSAY, 2 MAPS, 2 MAGAZINE COVERS

After entertaining the inevitable discussion of who deserved to win the 2000 U.S. presidential election, try moving into Andrew Sullivan's essay "Two Nations, Undivided" with a brainstorming exercise. The second paragraph ends with a striking list of opposites meant to be taken as descriptors for liberals and conservatives, stressing the ideological gap between them. Sullivan writes, "The divide can be described in a million ways: ethnic minorities versus white traditionalists; rootless cosmopolitans versus redneck bigots; Salon.com versus the Drudge Report; iMac users versus tow-truck drivers; Starbucks versus Cracker Barrel; NPR versus Rush." First begin by discussing the assumptions behind such a list—what is Sullivan implying about the class and educational backgrounds of these groups? Is Sullivan's view biased in any way? Next, have your class generate additions to the list, and follow this exercise with a discussion about our country's tendency to create a stable distance between political viewpoints.

You might then turn to the rest of Sullivan's essay, the purpose of which is to chip away at this false sense of divisiveness. To accomplish this purpose, the author first characterizes the divide by playing on our common stereotypes of Republicans and Democrats. He also mentions how geographically distinct the outcome of our past election really was—the final results showed a nation with liberal coastlines and a conservative center. But when Sullivan brings the reader into the smaller, more localized details, we see a nation that is not so split. States that went Republican also had a strong Democratic showing, and vice versa. He mentions the notorious lack of distinctiveness that characterized our presidential frontrunners and how that contributed to a tight race. And he eventually hones his argument to the level of the person, referring to political pundits like Mike Barnicle, Paul Begala, and George Stephanopoulos and noting how their personal backgrounds betray their current political stances. He concludes that the election signified that this illusion of divisiveness, which suits politicians more than it does society at large, is on the way out—to him, this is a good thing.

In examining the two *Time* covers, students should notice the use of cropping and computer-aided manipulation in each. What message is each magazine sending out to its readers, and what political stance (if any) are they taking? Examine the contents of these respective magazines to gather insights about their editorial alignment.

METHOD

The geographic divide that characterized the election not only parallels the nation's political divide but also falls along differences in race, class, spirituality, education, and level of diversity. These differences are characterized in Sullivan's catalog of liberal and conservative characteristics.

MESSAGE

Students should note that, according to Sullivan, Bush should have been the victor because he represents the massive red space on the second map. The population inhabiting this space also happens to be predominantly white, which suggests that multiculturalism may not have won out after all.

MEDIUM

In covers of this type, design decisions such as right or left and red or blue become politicized; positions and colors are metaphorical shorthand by which we recognize the candidates' political alliances. Though the first cover doesn't make an overt attempt to favor one candidate over the other, it might be noted that the stress placed on their similarities might have been seen as insulting to Gore, who was running for office while his was the party in power. Also, because there is no mention of a third party candidate, it can be assumed that *Time* is targeted at a primarily moderate audience accepting of the two-party stystem.

ADDITIONAL WRITING TOPICS

1. In a persuasive essay, argue for your particular choice of president. Who should have won the 2000 election? Consider as well all of the post-election controversy surrounding the uncounted Florida ballots, the Supreme Court decision, and any other factors you feel were important in this outcome.
2. In your opinion, why was the 2000 presidential election so close? Was it because of the types of issues the candidates ran on, their similar positions, their respective images, voter apathy, or some other set of factors? Explain.
3. You are running for president in 2004. Research political speeches in your library or online, and compose a mock campaign speech in which you outline your political position, your specific platform, and your impression of the last election. Make your speech compelling—you want people to vote for you.

DIVERGENCES

Print

Brinkley, Douglas, et al. *36 Days: The Complete Chronicle of the 2000 Presidential Election Crisis*. New York: Times Books, 2001. Collection of articles and reflective pieces on the election by various writers for the *New York Times*.

Sullivan, Andrew. *Virtually Normal: An Argument about Homosexuality and Love Undetectable*. New York: A. Knopf, 1995.

Web

http://www.msnbc.com/news/NW-front_Front.asp. The *Newsweek* and MSNBC partner site contains archived material on the 2000 election.

http://www.time.com. *Time* Online also contains archived election articles.

COLORS India

STORY, 3 PHOTOGRAPHS

Jhumpa Lahiri's "This Blessed House" is framed in the third person but is closely aligned with the character Sanjeev, whose thoughts are the ones to which the reader is most privy. (This is a technique called selective omniscience.) Sanjeev, a conservative and career-oriented young Indian man, is something of a curmudgeon compared to his wife, Twinkle, and his irritation at her hoarding of Christian memorabilia left over from the previous residents of their house is the story's main source of tension. At least on the surface this is the case—flashbacks into the couple's short past let the reader infer that the irritation has something to do with their hasty courtship and marriage. Sanjeev is aware that he barely knows Twinkle and her idiosyncrasies and that his decision to marry was in part due to cultural pressures. The story resolves after a dinner party in which Twinkle, who wins over Sanjeev's coworkers and colleagues, finds a silver bust of Christ in the attic—Sanjeev knows he can't win the battle over the tacky Christian trinkets displayed throughout his house, and decides to quietly carry on.

Ultimately, Lahiri's story gives readers a glimpse into Indian culture that is neither one-sided nor dependent on stereotypes for its characterizations—these feel like honest people, each distinct from the other. One stylistic technique the author employs throughout is the abundant use of sensory details. These details go beyond sight and sound—touch, smell, and taste are well represented throughout, and they help expand our understanding of Indian culture. To help your class make its way into this text, consider breaking your students into five groups, each group charged with the task of locating sections throughout the story that pertain to one of the five senses.

You could also compare Lahiri's story to Xavier Zimbardo's photography in terms of how each artist tries to represent Indian culture through a different medium. How does Zimbardo, for instance, use color differently than Lahiri? Do the characters in Lahiri's story, which, to an extent, the reader is forced to imagine, differ from the figures in the photographs? How does Zimbardo suggest to the viewer the role of spirituality in many aspects of Indian culture?

CALLOUT QUESTION

Obviously, responses to this question will vary, but you should explain to your students that the exercise is designed so that they accomplish more than simply changing the pronoun references throughout the paragraph—with a shift in point of view also comes a shift in how a certain character experiences (or is capable of experiencing) the world. They should note also the attempts at characterization that already exist in the paragraph—most of these references give us insight into Sanjeev's thoughts and motivations, which is fairly consistent with the rest of the story, too. For instance, the paragraph tells us that Sanjeev is "puzzled" and even "irritated" by the tackiness of the religious icons, as well as Twinkle's odd fascination with them. Students should ask themselves, if Sanjeev were the voice of the story, how would he convey these feelings to the reader beyond merely stating that they exist?

MESSAGE

On the surface, Sanjeev's reason for not wanting to display the memorabilia is religious, but the selective omniscience of the narrator gives us insights into his thought process. In part, Sanjeev resents that Twinkle is able to see some kind of unnamed value in these items; he almost envies her childlike ability to marvel at these things. Twinkle's motivations, in contrast, seem genuine—she finds the pieces fascinating because of her imagined empathetic connection with the previous owner. Then again, the story's point of view does not let us into her thoughts as easily as Sanjeev's, so we can't be entirely sure.

METHOD

Students should be able to describe Sanjeev as reserved, grumpy, practical, conservative, and consumed with his social position (perhaps the author's attempt to characterize the Indian American stereotype), while Twinkle is free spirited, childlike, and emotionally driven. The resolution is only implied, and is not complete by the story's end, though the fact that Sanjeev carries the bust of Christ for his wife without complaint suggests that their relationship is becoming solid.

MEDIUM

It may be necessary to refer to the captions included with Xavier Zimbardo's photographs in order to infer that the Indian culture is deeply invested in the Hindu religion—secular spaces are not as separate in this culture as they are in American culture. Asking students to identify the piece that struck them more immediately could be an effective way to begin a discussion aimed at comparing the two pieces.

ADDITIONAL WRITING TOPICS

1. You might consider your home as a refuge from culture at large. In a short freewriting exercise, describe the private life that you live in your home, the relationship you have with the people who live with you, and explain how that differs from the public life you lead outside your house.
2. In a first-person narrative, assume the point of view of the previous occupant(s) of the house in Jhumpa Lahiri's short story. Refer to the details that offer you clues about who might have lived there before, and provide a characterization of this person—his or her thoughts, beliefs, lifestyle, and the sort of activities that might transpire in a typical day.
3. Choose one of Xavier Zimbardo's pictures. First, describe it, and then write a brief narrative tying it to the other ones. Give details about the visual composition of the photo as well as your own personal reaction to it.

DIVERGENCES

Print

Lahiri, Jhumpa. *Interpreter of Maladies*. New York: Meridian, 2000.

Zimbardo, Xavier. *India Holy Song*. New York: St. Martin's Press, 2000.

Web

http://www.sscnet.ucla.edu/southasia/Culture/Cinema/cinema.html. The Manas site, which focuses on South Asian culture in general, offers several good articles on Indian culture, particularly India's rich contributions to cinematic history.

Audio/Visual

Chutney Popcorn. Directed by Nisha Ganatra. Starring Ganatra, Jill Hennessy, Madhur Jaffrey, Sakina Jaffrey. Rated PG. Mata Productions, Inc., 1999. Videocassette. This acclaimed film directed by Indian American filmmaker Ganatra offers an unusual and humorous perspective on both lesbian and Indian American communities, featuring an interracial lesbian parenting couple.

6 Redefining Media

The main themes of this chapter are celebrity, "the media," and art. You may want to introduce these themes to your students before they begin reading the chapter, asking them to find connections among them. It could be argued, for example, that all three are intimately related, and one could not exist without the other. In many cases, the popular media, through repeated exposure, create a celebrity who is well known only for being well known. Have students list examples of cases where this is true and cases where it is not true. Is it true for Princess Diana? She was propelled to celebrity through her royal marriage, and the popular media sustained that celebrity, though she did not actually create it herself by any artistic means. This also raises a bigger question: What does the word "celebrity" mean? How do your students define it? What about "media" or "art"? It is important to have students question these concepts and search for their meaning. A discussion of these concepts could give students an advantage when approaching the selections in this chapter—you may want to stress to them that, in order to make their arguments more cogent, they should constantly question the accepted meaning of words.

CELEBRITY
High Profile

ESSAY, SCULPTURE, MAGAZINE COVER

One engaging way to begin thinking about the pieces in this selection is to have your students (through collaborative brainstorming or individual freewriting) construct a list of current celebrities: singers, actors, comedians, royals, politicians, and other high-profile persons. Then have them vote on the "star power" of the various celebs—who's hot, who's not, who's dangerously overexposed, who's the next "it," who has achieved bona fide icon status. Follow up with a discussion of the criteria that make celebrities famous—physical attributes, personality traits, associations with other stars, and so on. Use these criteria as a yardstick to measure the specific celebrities that appear throughout this selection—Princess Diana, Michael Jackson, Julia Roberts, Brad Pitt, and Winona Ryder, among others. This activity is meant to illustrate to your students how genres are formed and defined—in this case, the celebrity type itself.

Katie Roiphe's "Profiles Encouraged" is a rhetorical analysis of a specific genre of journalistic writing, the celebrity profile. She defines the genre by listing the traits common to many of these profiles, making the eventual claim that because so many of them are identical—not only structurally but also at the level of what metaphors and adjectives are used—they fulfill a particular purpose for the audience. That we continue to read clichéd descriptions of celebrities blushing, or being vulnerable, or holding on to their childlike charm is not because we desire to get to know the real person behind the celebrity, but because we want to reaffirm our notions of fame. Roiphe closes her essay by saying, "True mystery doesn't interest us; the statement 'she has an aura of mystery' does. The clichés are what we crave and continue to expect. What makes glamour, like lights on a marquee, is the repetition of the familiar sounds of adoration, the same babble of fawning irony, the same vulnerable perfect creature we don't really want to read about." Encourage students to find celebrity profiles other than the ones Roiphe cites and compare their structure and figurative language to the archetype she defines. Is Roiphe's definition of the genre a fair one, or has she been selective in choosing her evidence in order to prove her point?

Finally, direct students to examine the visual texts in this selection. What sort of image does the *Time* cover try to project for Princess Diana? The shot is a close-up detailing Diana's smile, expensive earrings, and royal white teeth. Does this image confirm the generic attributes identified in Roiphe's essay? The comment accompanying Jeff Koons's sculpture states that "*Michael Jackson and Bubbles* [is] a work that comments not only on Jackson's unique celebrity, but on the very issues of identity and idolatry." You might discuss with your students Michael Jackson's "unique celebrity"—highly reclusive, given to purchasing bizarre items like the Elephant Man's skeleton, and occasionally accused of pedophilic crimes, Jackson is certainly atypical of the "common" celebs mentioned by Roiphe. Why might Koons have chosen Jackson as his subject over a less controversial celebrity?

MESSAGE

The visual texts in this selection could be read as making two different statements about fame, both of which are accounted for in Roiphe's essay. Koons's statue, for instance, comments ironically on what Roiphe, in her final paragraph, indicates is our real fascination with the celebrity profile genre: "Because, in the end, we are not interested in Winona Ryder; we are interested in fame: its pure, bright, disembodied effervescence." The statue of Michael Jackson and his chimp friend Bubbles embodies the idea of fame surrounding Jackson, a pop icon whose status is nearly godlike (and hence pure, bright, and disembodied). One of the generic conventions Roiphe identifies as "a vague way of satisfying the need for the movie star to be 'human' without detracting from her glamour with undue specificity" is the trait of vulnerability. On the *Time* cover, we see Princess Diana in extreme close-up, showing the viewer her "human," less stately side, which suggests at least a degree of vulnerability.

METHOD

Roiphe's argument is supported by several examples for each point she brings up; however, students might notice that she tends to cite the same profiles again and again—profiles on Gwyneth Paltrow, Julia Roberts, Brad Pitt, Winona Ryder, and Tom Cruise serve as supports throughout the piece. The question of how effectively this repetition works is, of course, a matter for debate and depends on what sort of audience reads the piece. Though some might think the limited examples have been hand-picked in order to artificially enhance the writer's stance, others might argue that extensive examples are not needed, especially if the reader already has some familiarity with the genre and accepts Roiphe's conventions as accurate in the first place.

MEDIUM

By using subject matter from popular culture, Koons has violated the distinction between high and low art. Further, the style of the piece suggests religious iconography (think of the small ceramic icons of Christ and the Virgin Mary associated with Catholicism, for instance) and thus puts the pop culture figure of Michael Jackson into yet another context. This postmodern gesture of Koons's indicates how pervasive the middle class has become in reshaping our notions about art in general. Finally, a nineteenth-century way of thinking about artistic creation was that art came solely from the artist's imagination; postmodernism, in contrast, borrows from the images and symbolic associations of society at large, combining them in a mishmash sometimes called "pastiche" art.

ADDITIONAL WRITING TOPICS

1. In a fifteen-minute freewriting exercise, give your personal reaction to Princess Diana's death. Were you at all affected by it? Do you have an opinion on whose fault the tragedy was? Were you surprised by the news and the subsequent public reaction to the death?
2. Using the generic conventions identified in Katie Roiphe's essay, write a mock celebrity profile on a star of your own choosing. Follow up this piece with a brief reflective statement

that points out to the reader how you incorporated these various celebrity profile clichés in your own writing.

3. Spend some time researching celebrities from the last fifty years or so. Compare the images of earlier celebrities to the images of celebrities now. How, for instance, do we see Eminem differently than fans of Elvis saw him? In what ways are Madonna and Marilyn Monroe alike? Use several examples in your essay.

DIVERGENCES

Print

Ellis, John. "Stars as a Cinematic Phenomenon." In *Film Theory and Criticism: Introductory Readings*. Edited by Leo Braudy and Marshall Cohen. New York : Oxford University Press, 1999. This essay discusses film in terms of star theory, how a celebrity embodies a set of predetermined associations separate from and part of each film.

Roiphe, Katie. *Last Night in Paradise: Sex and Morals at the Century's End*. Boston: Little, Brown and Co., 1997.

———. *The Morning After: Sex, Fear, and Feminism on Campus*. Boston: Little, Brown and Co., 1993.

Web

http://www.artcyclopedia.com/artists/koons_jeff.html. Artcyclopedia has a list of links to online galleries and articles featuring postmodern artist Jeff Koons.

http://www.greatdreams.com/princess.htm. This fan site on Princess Diana features multitudes of links to information concerning her life, her celebrity, her tragic death, and the conspiracy theories associated with that tragedy.

Audio/Visual

Celebrity. Directed by Woody Allen. Starring Leonardo DiCaprio, Hank Azaria, Winona Ryder. 113 minutes, rated R. Mirimax, 1997. Videocassette, DVD. This characteristic black-and-white Woody Allen film satirizes the excess of celebrity culture and, ironically, features cameos by some of Hollywood's prominent stars.

FRONT PAGE
Elián Gonzáles

ESSAY, 3 NEWSPAPERS

William Saletan's "The Elián Pictures" is a critical analysis of the almost iconic images emanating from the Elián Gonzáles custody battle; it offers students a good model for deconstructing the claims of truth often merely assumed in photojournalism. The essay is structured around a series of simple questions that are used not to get to the truth of the matter but rather to point out how the truth has been obscured. Ironically, these are questions that journalists typically ask—who, what, when, where, why, and how. The essay thus provides students with an easily adaptable template for examining similar examples of news reports. You could construct an analytical exercise in which students choose other contentious texts from the news media (not confined only to news photography), devise their own Saletan-styled questions, and provide answers meant to show how such inherently persuasive texts are not in fact objective.

You might also consider having your students discuss Saletan's neutral stance. In showing his audience how one might be equally skeptical of the two photographs, Saletan is able to criticize each polarized position. But also ask whether Saletan really is an equal-opportunity critic or whether his language choices indicate a bias toward one side or the other. His main argument, however, finally has little to do with the specific crisis of a young Cuban boy in the midst of a diplomatic tug-of-war. Rather, his focus is on the news media's insistence that pictures such as these represent reality. These photographs provide a site of examination where Saletan can show how news images—and the stories told around them—are subject to political or personal manipulation.

This selection also presents a good opportunity for having your class discuss the print medium—specifically, newspapers. Ask students to compare the layouts of each front page and consider whether the various elements create an argumentative stance on the Elián controversy (as suggested in the Method question below, each newspaper has its own opinion about the issue). Point out compositional factors such as the relative size of photographs and headlines; the hierarchical placement of elements (for instance, what is placed on the right side of the page versus the left or above the fold versus below); and the positive or negative connotations behind the words chosen for the headlines. After identifying these various features, students should then be expected to explain how they help convey (albeit subtly) a particular editorial position to the reader.

MESSAGE

Taken together, the two photographs Saletan analyzes served to polarize the public opinion surrounding the Elián controversy. The seizure photograph shows the situation at its worst, an armed government agent in full riot gear snatching away a petrified child—we are meant to condemn the government's actions upon seeing this image. The second picture gives us the image of a happy child reunited with his family—here, we should feel that the proper course of action was taken. Saletan's argument is that these two images, both of which are purportedly objective, actually help create the divisiveness, the either/or positions, in the Elián situation. Other photographs that were used served to back up one side or the other. According to Saletan, this is how the news media usually operate when presenting issues of controversy to the American public. Consequently, he refuses to take a stand in his reading of the photographs. Instead, he complicates both positions by being sympathetic to the opposing side in each of his scrutinizing evaluations.

METHOD

The *Miami Herald*'s front page seems sensationalized, with the raid photograph taking up the largest portion of the above-fold area. Also the large-type headline "SEIZED" emphasizes the act itself, and adjectives in the related headlines—"angry," "tense," "shocking"—suggest that the paper's sympathy lies with those wanting Elián to stay in the United States (which is not surprising, given the high anti-Castro sentiment among its Cuban American readership). By contrast, the *New York Times* front page prominently shows the reunion photo, but not the high-drama moment of seizure. The language of the headline "REUNITED WITH HIS FATHER" suggests that the seizure was the right decision, a suggestion echoed in the government-sympathetic headline "For Reno, a Difficult Call in the Last Minutes." Of the three, the *Los Angeles Times* perhaps offers the most evenhanded treatment of the story, with the two side-by-side photographs taking up roughly equal amounts of space.

MEDIUM

Saletan's critique has nothing to do with the technical expertise of the photography; its subject, rather, is the illusion that news photography tries to perpetuate. Namely, Saletan takes issue with the notion that these photographs are offered to the public as if they were entirely objective, factual records of the news event, putting us in the moment as it actually occurred. In actuality, the photos are the subjective products of human beings with vested interests in the issue at hand, and elements like cropping, perspective, and lighting are often results of that subjectivity. Though he does not suggest outright an alternative that might lead us out of this trap, Saletan wants us to be more adept at scrutinizing news photography; he wants us to learn to ask the right questions so that we no longer think about complex, controversial issues in simplistic, black-and-white terms.

ADDITIONAL WRITING TOPICS

1. William Saletan's "The Elián Pictures" is structured around a series of specific questions designed to challenge the apparent objectivity of two news photographs. Choose another set of memorable news photos, video footage, or sound bites and adapt Saletan's questions in an analysis of your own.
2. After being given a random photograph by your instructor, construct two contrasting news scenarios in which the photograph could be used. Try to make both stories equally plausible even as they oppose each other in terms of content.
3. Research the history of diplomatic relations between the United States and Cuba. Using your research as evidence, write an essay in which you speculate on how Cuban news coverage of the Elián Gonzáles custody battle would differ from the U.S. coverage. Think in terms of form as well as content, and address how the two images included in this selection might be explained by Cuban news officials.

DIVERGENCES

Print

Cohen, Herbert, ed. *125 Years of Famous Pages from* The New York Times: *1851–1976*. New York: Arno Press, 1976. This collection offers a wonderful illustration of how one of America's top papers has become more visually oriented over the years.

Web

http://slate.msn.com/. William Saletan's article originally appeared in *Slate*, the Microsoft-sponsored online politics and culture magazine.

http://www.moviejuice.com/elian/moviereport.htm. This spoof site features "Elián" offering his opinions of recent Hollywood films. Explore the cultural associations behind this particular brand of humor.

http://www.pathtofreedom.com/main.html. The "Path to Freedom" Web site was founded as part of a movement of those in favor of Elián's return to Cuba; it can be used in conjunction with analyses of newspaper layouts.

BLAME
Media and Violence

POSTER, 2 ESSAYS, 4 FILM STILLS, 3 COMICS

The two essays in this selection offer clear, contrasting examples for illustrating certain rhetorical concepts. First, what do students know about Marilyn Manson? Are they less likely to agree with Manson's points because of his inflammatory stage persona? Conversely, are they familiar with Gerard Jones? Whether they are or not, his status as the author of violent comic books stands as an interesting counterpoint to Manson's role. Is one more likely to agree or disagree with Manson based on his career as a provocateur? Is Gerard Jones's commentary valuable because of his role in media violence, or is he part of the problem? Would Manson agree that violence is, in fact, a necessary outlet for young boys, or would he decry Jones's work as a perpetuation of society's already violent tendencies? Also, have students consider how each essay is written—in terms of style, Jones's piece is much more formal and argumentative in structure and tone, whereas Manson's defense is more casual, free-associative, and loosely structured. Based upon these stylistic choices, ask students to define what audience each writer imagines himself addressing—one that is already in agreement, or one that needs to be convinced? The fact that Jones' article initially appeared in the progressive magazine *Mother Jones*, while Manson's appeared in the popular music magazine *Rolling Stone* gives us at least an initial hint as to what these different audiences might look like in terms of age, class, cultural positioning, and political bent.

Students should be familiar with *Tomb Raider*, *Natural Born Killers,* and *Pulp Fiction*, but ask them if they have ever seen *Bonnie and Clyde*. Are they surprised that such a movie appeared more than twenty-five years ago? Also, have students consider the three Gerard Jones comics. Are they just fantasy, do they serve Jones's goal of providing an outlet for kids' aggression, or are they part of the problem of violence?

MESSAGE

Gerard Jones establishes blame on the part of a culture that teaches children "early on to fear our own" rage. Manson sees the blame as much more identifiable—in his opinion, society is inherently violent, and its religious and cultural traditions are steeped in glorifying and aestheticizing violence.

METHOD

The main thrust of Jones' essay is that "identification with a rebellious, even destructive, hero helps children learn to push back against a modern culture that cultivates fear and reaches dependency." He supports this statement with fairly compelling examples including his son, a girl who was "exploding with fantasies," and a woman he knew in college, among others. Manson theorizes more than he makes concrete claims, describing how his music offers hope and a voice to millions of alienated teenagers. He relies less on examples and more on his ideas for why society is so violent.

MEDIUM

Encourage students to consider the main idea of *Convergences*, that is, that the medium can sometimes be the message. For instance, Gerard Jones's comics, because they rely so heavily on the interplay between image and text, end being something of a synthesis between violence and the commentary on violence. Oliver Stone can be argued to be doing the same in *Natural Born Killers* by satirizing the public's taste for violence.

ADDITIONAL WRITING TOPICS

1. Reflect on your past encounters with violent fiction—books, films, cartoons, comic books, video games, and so on. Describe three of your more striking experiences, as well as any aftereffects or lingering memories you have of these encounters.
2. Decide which essay in this selection evoked a stronger reaction from you, whether you were for or against the author's position. Write a letter to that author in which you either commend him for his views on the subject or express your disagreement with his opinion.
3. One long-standing debate in the world of art deals with freedom of expression versus social responsibility. In your view, how can this debate ever be reconciled?

DIVERGENCES

Print

Ellis, Brett Easton. *American Psycho*. New York: Vantage, 1992. Ellis's much-despised novel chronicles the episodic violent exploits of 1980s serial killer/Wall Street yuppie Patrick Bateman.

Ha, Gene, and Gerard Jones. *Oktane*. Dark Horse Comics, 1996.

Jacobs, Will, and Gerard Jones. *Tommy and the Monsters*.

Manson, Marilyn. *The Long Hard Road Out of Hell*. New York: Reagan Books, 1999. Manson's controversial autobiography includes an introduction by filmmaker David Lynch.

Web

http://www.aap.org/advocacy/mediamatters.htm. The "Media Matters" site is part of the American Academy of Pediatrics' National Media Education Campaign.

http:www.marilynmanson.net/. The official Marilyn Manson site will provide helpful background on this author's life and work.

Audio/Visual

Antichrist Superstar. Music and lyrics by Marilyn Manson. Interscope Records, 1996. Compact disc.

Bonnie and Clyde. Directed by Arthur Penn. Starring Warren Beatty, Faye Dunaway. Warner Bros., 1967/2001. Videocassette. 114 minutes, rated R.

Lara Croft—Tomb Raider. Directed by Simon West. Starring Angelina Jolie. Paramount Home Video, 2001. Videocassette. 100 minutes, rated PG-13.

Marilyn Manson. WHYY, Philadelphia, February 24, 1998. On this episode of National Public Radio's *Fresh Air*, host Terry Gross interviews the shock rocker in surprisingly civil fashion.

Natural Born Killers. Directed by Oliver Stone. Starring Woody Harrelson, Juliette Lewis. Vidmark/Trimark, 1997. Videocassette/DVD. 182 minutes, rated R.

TRUTH IN ADVERTISING
Jay Chiat

ESSAY, TV COMMERCIAL, ADVERTISEMENT

Some of the most striking media events of the last fifty years have been commercial advertisements, and a sure way to jumpstart your students' engagement with this selection is to have them reflect, in writing or in a general discussion, on their favorite commercials. Students' recall of campaigns such as "Fall into the Gap," Burger King's "Where's the Beef?", or "Enjoy Coca-Cola" should offer a peripheral lesson concerning how our shared popular culture is formed by such texts. As a class, develop a list of well-known commercials and identify what made them successful: specific imagery, catchy slogans, the use of music, celebrity spokespersons, and so on.

You might then consider shifting attention to Apple Computer's *1984*, thought by many to be the best television commercial ever made. Try showing the commercial in its entirety (several Quicktime versions of the video can be found on the Web; visit the URL listed below for one version). In discussing it, students should note how the commercial's creators used color, movement, costuming, and sound to contrast two opposing paradigms of technology—IBM's dismal clone-like world versus Apple's independent, free-thinking vision. (Of course, you might also point out the logical conflict inherent in the idea that buying any mass-produced item could promote individuality.) Next, focus on the other ads included in this selection: What is Apple suggesting when it links famous figures with its "Think Different" slogan? Is buying an Apple computer supposed to help us aspire to be like these visionaries, or does it suggest that the company is itself inspired by such figures? And why does the slogan use the adjective "Different" instead of the adverb "Differently"?

You could begin discussing Jay Chiat's "Illusions Are Forever" by asking students to identify how the author attempts to establish a trustworthy persona while presenting a challenging topic. Do they think his attempts are successful? In other words, do they trust that Chiat is telling the truth? Right away, he confronts the audience's suspicion that someone in advertising is the least qualified person to discuss the concept of truth by suggesting that those persons most opposed to a concept are the ones most aware of it. He then goes on to make a subtle distinction: commercials do not lie to audiences outright (most are factually accurate), but they do present viewers with unrealistic idealizations. Chiat's ethical concern is that the line between these conventional commercial messages and our sense of "personal truth" is being blurred by the effects of reality television shows like *Survivor* and the anonymous space of the

Internet—because they appear "unmediated," they disrupt our ability to separate fact from fiction. Ultimately, he sees new technology as a potentially democratizing force, able to "put CNN on the same plane with the freelance journalist and the lady down the street with a conspiracy theory." Engage students in a debate of Chiat's hopeful claim—do they share his optimism with regard to the Internet? You might also delve into a metaphysical discussion centered on the various definitions of truth invited by Chiat's last line, "After all, isn't personal truth the ultimate truth?" What things do your students hold true? How does commercial culture strive to manipulate such beliefs in order to sway its audience?

MESSAGE

The ad suggests that the Macintosh computer promotes freethinking, independence, and individuality. This stands in direct opposition to the IBM image of sheeplike conformity, represented by the commercial's anonymous crowd of lifeless people in dark blue suits. The commercial, of course, is a work of fiction, but as Chiat points out, in nearly every instance, associating a material product with nonmaterial values and beliefs is ultimately manipulative. He acknowledges the same phenomenon at work in his own advertising as well.

METHOD

Knowing the central theme of George Orwell's novel *1984*—an individual's struggle against an oppressive government—is key to fully understanding Apple's commercial. To some extent, literary allusions such as this work to send a subtle message to the viewer that the product is associated with a high-brow, learned image. However, since this theme is also fairly pervasive through literature and film, even those who are not familiar with the novel can "get" the commercial.

MEDIUM

Other than direct, lived, personal experience of the world around us, true unmediated events are a rarity. Challenge your students to list some of them, reminding them that in thought and speech a medium has been inserted into the equation, along with its characteristic way of distorting reality.

ADDITIONAL WRITING TOPICS

1. The *Advertising Age* quotation accompanying Jay Chiat's essay gives Apple's *1984* commercial credit for, among other things, "turn[ing] the Super Bowl into a major ad event." In a short descriptive essay, recall some of the Super Bowl ads that you remember from years past, explaining why they were so memorable to you.
2. Jay Chiat writes of the digital world: "But I believe technology, for all its weaknesses, will be our savior. The Internet is our only hope for true democratization, a truly populist forum, a mass communication tool completely accessible to individuals." Do you agree with Chiat's assessment? To what "weaknesses" is he referring? Defend or refute his position in your own argumentative essay.
3. Advertisements are designed to sell not only products but also images and associations. In your opinion, how far is too far when it comes to audience manipulation in advertising? Be sure to cite several examples to support your stance.

DIVERGENCES

Print

Orwell, George. *1984*. 1949. Reprint, New York: New American Library Classics, 1990.

Web

http://www.newspeakdictionary.com/. The Newspeak Dictionary site is based on the state-sanctioned language George Orwell developed in *1984*.

http://www.uiowa.edu/~commstud/adclass/1984_mac_ad.html. Sarah Stein's essay offers an adept reading of Apple's *1984* ad. This site also includes a Quicktime video version of the commercial in its entirety.

Audio/Visual

Crazy People. Directed by Tony Bill. Starring Dudley Moore. 90 minutes, rated PG. Paramount, 1990. Videocassette. This comedy centers on a highly successful advertising campaign created by the patients in a mental health institution; the slogans rely on plain, direct statements such as: "Volvo—Boxy, but Good."

CORPORATE JAMMING
Nike

ESSAY, ADVERTISEMENT, ANTI-AD, SCREEN SHOT

Begin exploring this selection by having students visit the Adbusters Web site so that they can see additional "anti-advertisements." You can then have them discuss why these spoof ads are funny—what intertextual references do they make, and what should a viewer know about the targeted companies in order to get the joke? Remind students that the purpose of these spoofs is not just to amuse. Adbusters has a specific agenda, which it calls "corporate jamming"—criticizing corporate practices by using the very language of commercialism (its slogans, images, and branding identities) against itself. In identifying such features, you might also discuss Nike's doubly parodic advertisement—is Nike's strategy of mocking the attempts of others to mock the company an effective one, or is it simply too ironic and detached?

Jonah Peretti's e-mail correspondence regarding the "Nike iD" campaign accomplishes with words what Adbusters' "uncommercials" do in multimedia spaces. Ask your students to characterize Peretti's tone—they should notice that he "borrows" Nike's commercialese, and thus the company's authority, as he tries to get his "sweatshop" custom shoe request honored. Ask students also to characterize the anonymous, measured tone of the Nike e-mails—what is such a tone supposed to suggest to readers about the relationship between customer and company? You might have students view the entire correspondence at the URL listed below, after which they can discuss the eventual outcome of the dialogue. Who seems to have "won"?

Finally, you should discuss the claim made in Peretti's article—that micromedia offers individuals a powerful tool for resisting corporate power. Do your students agree that this method of media control is an empowering one for the little folks? Or do they think it offers an easy opportunity for greedy companies and scam artists to inundate people with their sales pitches? Have students find examples of Internet-based micromedia actions, citing both beneficial and detrimental instances: e-mail petitions, chain letters, junk mail, the popular and widely disseminated URL to the "All Your Base Are Belong to Us" Web site.

MESSAGE

Like Jay Chiat, Jonah Peretti believes that the Internet has the potential to revolutionize how individuals receive information about the world around them: it lets them resist the pressures brought on by forces like commercialism, the entertainment industry, and mass media. As an advertising professional, Chiat's ultimate goal is not to subvert or overthrow commercial culture and mass media, but rather to call for a better understanding of how these entities use certain strategies to manipulate their audiences. In contrast, Peretti, if not revolutionary, is certainly taking corporate culture to task for what he sees as unethical practices.

METHOD

Peretti advocates "micromedia" in contrast to the uncaring, tired content of the mass media. He envisions a bottom-up network of distributing information that relies heavily on the Internet and other digital modes of communication. His own actions imply that he thinks this method could be employed by anyone.

MEDIUM

It is likely that students will point out the fact that network television is primarily a commercial enterprise. The networks will not run Adbusters' spoofs in part because they do not wish to alienate companies on which they rely for advertising revenue.

ADDITIONAL WRITING TOPICS

1. Look up the word "parody" in a dictionary. In a short essay, describe what features of Adbusters' "uncommercials" make them parodies. What assumptions about commercialism and corporate ethics are being attacked in these fake ads? Be sure to cite specific examples in your definition.
2. The Adbusters position statement draws a parallel between the organization's practice of "culture jamming" and the 1960s civil rights movement, the 1970s feminist movement, and the 1980s environmental movement. Do you agree with this assessment? Compare and contrast the work of Adbusters to that of these historic activist movements.
3. Working with a partner, choose a persona and continue the e-mail correspondence between Jonah Peretti and the customer service representative from Nike. Before writing, discuss with each other how you will employ the writing styles and rhetorical techniques of both writers' e-mails.

DIVERGENCES

Print

Frank, Thomas, and Matt Weiland, eds. *Commodify Your Dissent: Salvos from* The Baffler. New York: W. W. Norton, 1997. This collection of essays from the anticorporate journal *The Baffler* deals with how corporate America transforms rebellion against commercial culture into a commercial product itself.

Lasn, Kalle. *Culture Jam: The Uncooling of America*. New York: HarperCollins, 2001. The editor of *Adbusters* discusses the ubiquitous branding of American culture, arguing that there is no essential difference between our national identity and our identity influenced by corporations.

Web

http://www.adbusters.org/. The Adbusters Web site is an essential resource for discussing this selection.

http://www.nikeid.com. On the Nike iD site, visitors can still order customized running shoes.

http://www.shey.net/niked.html. This site includes the original e-mail correspondence between Jonah Peretti and Nike, as well as links to other articles.

BLURRED LINES
Poetry and Painting

POEM, 2 PAINTINGS

Read Frank O'Hara's poem "Why I Am Not a Painter" aloud in order to give the class an opportunity to experience it as a performance. Then, explicate the poem together. You might start with basic questions about the events outlined in the poem or any images and phrases that the students find striking—in short, what is their knee-jerk, emotional reaction to the poem? This could naturally lead into talk about the intended mood, which might be described as a sort of casual frustration, as well as a characterization of the poem's speaker—what kind of personality traits do we find in this speaker? What hobbies might he have other than poetry? How does he feel toward his chosen craft of writing as opposed to that of painting? Encourage your students to be imaginative in constructing a portrait of O'Hara's persona.

You can connect the poem to the paintings in this selection by discussing the similar processes involved in creating art. O'Hara draws a parallel between the poet's journey to create a poem about a color and artist Mike Goldberg's journey to create a painting based on the word "SARDINES"—neither journey carries the artist to his expected destination. Though painting and writing are different media, the creative processes of selecting, reflecting on, and transforming a subject into a truly original piece of work are analogous, and you may want to extend O'Hara's analogy to include other media. To connect this selection to your students' own writing, you might initiate a conversation about the various approaches your students take when composing their essays, letters, and so on.

Additionally, both Mike Goldberg's and Jasper Johns's works incorporate a verbal dimension into their finished product, much as O'Hara's poem uses painting as its controlling metaphor. Ask your class if the allusions to other media suggest which medium is better at accomplishing certain goals.

MESSAGE

Though O'Hara's poem *refers* to Goldberg's painting, it does not *describe* it. In fact, the only concrete detail O'Hara includes is that at one time the painting contained the word "SARDINES." The accompanying comment by literary critic Anthony Libby succinctly states O'Hara's goal: "What replaces image is not so much plot, though the poem does tell a story and imply an argument, as the movement of the individual line, frequently circling back on itself to create the duration as well as the depth and texture of a particular experience." A reader might construe that O'Hara envies Goldberg's freedom as a painter, his ability to transform a word into something more than (or no longer) a word. Within the medium of poetry, O'Hara cannot exactly turn the word "orange" into a nonlinguistic image—he is constrained by the medium of language.

METHOD

O'Hara's language is not stereotypically poetic, full of lofty and ornate diction; in fact, we recognize it as poetry only because of its use of line breaks and spacing. The ordinariness of diction is in keeping with the subject matter—the poet wishes to an extent that he were able to work from within the medium of painting. The poem is self-reflexive as well; it becomes in the final stanza an examination of the speaker's own experience with writing poetry (seen as the equivalent of Goldberg's experience producing a painting) and how that experience organically changes during the creative process.

MEDIUM

Painter and poet alike start out with a vision of a completed work. Mike Goldberg initially chooses to include "SARDINES" in his painting because, as he suggests in the poem, "it needed something there." The poem's speaker sets out to write about the color orange. By the end of each act, Goldberg has gotten rid of his initial vision and reduced the word to indecipherable letters; O'Hara's persona has written no less than twelve poems, none of which mention the color orange. All of the pieces in this selection, including Johns's *False Start*, deal with verbal language (or the actual letters in the alphabet) in terms of its graphical or aesthetic qualities.

ADDITIONAL WRITING TOPICS

1. After studying O'Hara's poem, write a short prose reflection on a medium other than writing: painting, film, music, or whatever else you find interesting. Do not focus on specific works within your chosen medium, but rather on the medium itself.
2. What is your aesthetic reaction to the two paintings in this selection? That is, which one (if either) do you like better, and why? Describe their elements—such as color, line, form, and overall composition—and tell whether or not you find these elements visually appealing, confusing, or distasteful.
3. Just as O'Hara's poem muses on the word "ORANGE," Goldberg's painting initially plays with the word "SARDINES." In a ten-minute writing exercise, try constructing your own poetic reflection on these words.

DIVERGENCES

Print

O'Hara, Frank. *The Collected Poems of Frank O'Hara*. Edited by Donald Allen, with an introduction by John Ashbery. New York: Knopf, 1971.

Web

http://wings.buffalo.edu/cas/english/faculty/conte/syllabi/377/Frank_O'Hara.html. This University of Buffalo faculty member's site features Frank O'Hara's poem along with artwork by Jasper Johns, Larry Rivers, and Mike Goldberg.

http://www.artchive.com/artchive/J/johnsbio.html. The Artchive site features an extensively hyperlinked essay on Jasper Johns's work in the context of the Pop Art movement.

http://www.english.uiuc.edu/maps/poets/m_r/ohara/ohara.htm. Modern American Poetry's entry on Frank O'Hara also contains links to several good short articles on the poet.

Audio/Visual

The Simpsons. "Mom and Pop Art." 23 minutes. Fox Broadcasting, April 11, 1999. Homer becomes an instant celebrity in the art circuit thanks to his rage-inspired "outsider art." The episode features cameos by Jasper Johns and Isabella Rossellini.

MEDIATION
Art and Nature

ESSAY, INSTALLATION

Terry Tempest Williams's "A Shark in the Mind of One Contemplating Wilderness" can be used in conjunction with a discussion about the related roles of form and style in writing. As an essay, this piece stretches the conventional boundaries of the form. By incorporating other texts in such a way that they are actually part of the essay instead of just supporting quotations, Williams composes (or compiles?) a truly multivocal text. Ultimately, the piece is a meditation on how to repair our culture's disaffection for nature by redefining nature as art—this meditation is brought on by Williams's multiple experiences of the shark in cultural and artistic settings. Williams's own musings are combined here with Damien Hirst's interview and aphoristic blurbs from Thomas McEvilley and Federico García Lorca to create an almost hypertextual, associative space that is far more open-ended than students might expect of essayistic prose. Ask your students what they think of this patchwork style—how do they conceive of the author's presence in a piece that imposes no controls on the other voices it utilizes? Might this be a style that they would want to incorporate into their own writing?

A discussion of Hirst's installation *The Physical Impossibility of Death in the Mind of Someone Living* could be used as a lead-in for your students as they write about other works of art. First of all, you should ask your students if they think of Hirst's installation as legitimate art—if not, how might they redefine it? They should attempt an explanation of what effect Hirst's work has on viewers—beyond just shocking them, is the work intended to tease out our inability to properly imagine what death is like? You should ask them to consider the metaphysical implications behind the piece's title and how the subject matter reflects that title. Draw their attention also to the choice of medium, as well as the materials used in the construction of the piece. These general considerations, among others, should be taken into account when analyzing any work of art.

MESSAGE

As the metaphor in her essay's title implies, Terry Tempest Williams is writing in a manner that mimics the shark's movement—darting all over the place in an effort to show her readers how her "mind becomes wild in the presence of [artistic] creation." The form of the essay, then, echoes its function, which is to contemplate nature as a work of art. Thinking along these lines, Williams juxtaposes descriptions of the natural world and excerpts from interviews with installation artist Damien Hirst. Williams brings urgency to the essay by suggesting that people find value in the wilderness by finding value in art.

METHOD

Each of the paragraphs in Williams's essay is a description of a shark in a different context: in an aquarium, a natural history museum, and an art museum. Williams is in each scene a passive observer who philosophically reflects on these experiences and her conception of the shark. In every scene, she is able to eventually imagine the shark as a living, moving being. This series of moments illustrates her ultimate argument—reconceiving of wilderness as art as a possible means of ensuring its preservation.

MEDIUM

You might want to ask students to think about Damien Hirst's decision to tackle the issue of death—what questions does his installation raise about the way we think of our own mortality?

ADDITIONAL WRITING TOPICS

1. How do you define the words "nature" and "culture"? Are they distinct and separate concepts, or do you find spaces and moments where the two are combined or entangled? In a fifteen-minute freewriting exercise, try to construct definitions based on any previous associations you have had with these terms.
2. Using Williams's essay as a frame of reference, write an essay in which you describe in detail a piece of art you find provocative or pleasing. Try to explain your reaction to the piece as well—is it the form or subject matter that causes you to feel the way you do? You might consider visiting a nearby museum or consulting an art book for inspiration.
3. One of the concerns expressed in Williams's essay is the ever-encroaching presence of civilization on nature; in her last paragraph, she writes, "The natural world is becoming invisible, appearing only as a backdrop for our own human dramas and catastrophes." Do you agree with Williams's sentiment? Can you think of instances when civilization becomes invisible, when the natural world becomes more than mere decoration?

DIVERGENCES

Print

Morgan, Stuart, ed. *Damien Hirst: No Sense of Absolute Corruption*. London: Gagosian Gallery, 1996.

Williams, Terry Tempest. *Coyote's Canyon*. New York: Gibbs Smith, 1999.

———. *Desert Quartet*. New York: Pantheon Books, 1995.

———. *Leap*. New York: Pantheon Books, 2000.

———. *Pieces of a White Shell: A Journey to Navajoland*. New York: Scribner, 1984.

———. *Refuge: An Unnatural History of Family and Place*. New York: Pantheon Books, 1991.

———. *An Unspoken Hunger: Stories from the Field*. New York: Pantheon Books, 1994.

Web

http://abcnews.go.com/sections/world/DailyNews/germany010329_body.html. This URL leads directly to the ABC News.com article on German artist Gunther von Hagen's controversial exhibition *Body Worlds*, which features preserved human corpses in various poses.

http://www.apple.com/applemasters/dhirst/. Apple Computer's "Think Different" online campaign names Damien Hirst as one of its Applemasters and gives a short bio statement. This Web site can be used in conjunction with the "Truth in Advertising" selection earlier in this chapter.

Audio/Visual

Salt Lake City Remote: The Changing West. National Public Radio, May 14, 1998. Audiocassette. *Talk of the Nation* host Ray Suarez interviews Terry Tempest Williams and others about the changing landscape of the American West from a pristine environment into one full of strip malls, subdivisions, and ski resorts. Available through http://www.npr.org.

TV ART
Nam June Paik

2 INSTALLATIONS

The installations of Nam June Paik included in this selection tackle the project of representing a particular medium itself as art and thereby highlighting the hidden aesthetic qualities of that medium. Begin with the quote from Guggenheim curator John G. Hanhart in the introduction to the selection: "The artist stepped into the television, broke the border of the frame, and . . . reinterpreted video and television as a visual and auditory instrument." Each of the pieces by Nam June Paik included here works from a similar artistic idea, the surprising juxtaposition of video technology and "natural" environments. In *Video Fish* we see an unnatural combination of TV monitors and living fish; the actual content on the monitors is diminished at the expense of concentrating on the purely aesthetic experience of video images filtered and refracted by yet another screen—an aquarium filled with water. *Megatron/Matrix* brings the television out of its intimate, cloistered living room and displays it as a sublime, almost sacred monolith—a wall of televisions with different large-scale images and an accompanying soundtrack.

In directing your students to think about the ways that Paik decontextualizes our relationship to television so that we see it as both a medium through which content is relayed as well as an artistic object in its own right, you should start off by having them talk about our culture's conventional relationship to the medium of television. You might consider asking them questions such as these: Where do we usually see televisions, in both public and private spaces? What are the different purposes served by television? Does it only entertain, or can it also educate, distract, or placate? How does television influence our culture at large—do we find ourselves borrowing television's particular discourse in other spheres of communication? Answering these questions should establish a working definition for the traditional video context that Paik's work unsettles—then you should discuss what purpose this unsettling serves. Is Paik saying that we should simply be more aware of the aesthetic qualities of video as an artistic object? Or is he making some sort of politically charged statement about the ever-present (and perhaps detrimental) influence of television in every facet of our lives?

MESSAGE

According to Marshall McLuhan, there are no unmediated events beyond the realm of pure thought or experience—at base, translating an idea into language is to put that idea into a medium. In a general sense, Paik's artistic statement involves pointing out the interrelatedness of art, nature, and new media technology.

METHOD

Video Fish is like other Paik installations in that it blends a natural, immediate environment with the technological distance of the video monitor (a technology made to show us things that are not in our world). Also, there is a formal similarity between the shimmering waves and the play of light in water and the wavering images emanating from the monitors. Though the images of fish might serve to emphasize the similarities between these two media, the images of dancing and planes give the piece a surreal feel.

MEDIUM

Unlike the usual notion of sculptures, Paik's installations are dynamic—they move through space and time, constantly re-creating themselves. Seeing a static photograph of the work is definitely not the same as seeing it in person, where you get the opportunity to fully experience this movement. This loss of artistic energy may not be as obvious when discussing a static work of art such as a painting, but some art theorists argue that the reproduction of art in books and prints can only create inferior copies that lack the inspiring "aura" of the original.

ADDITIONAL WRITING TOPICS

1. Reflect on your earliest experiences with the media used in Paik's installations. What memories do you have, for instance, of your family's interaction with the television set? What shows did you watch as a child?
2. In a collaborative brainstorming session, try to imagine other video monitor installations that would imitate Paik's work, keeping in mind the artist's desire to accentuate the blending of nature and technology. You may include just a verbal description or, if necessary, rough sketches.
3. Guggenheim director Thomas Krens said in reference to the influence of new media on contemporary culture, "The challenge for museums is to bring this history of media and contemporary practice into their exhibition and collection programs." Imagining that you are a museum curator, design a proposal for an exhibition that addresses this concern.

DIVERGENCES

Print

Benjamin, Walter. "The Work of Art in the Age of Mechanical Reproduction." In *Illuminations: Essays and Reflections*. Edited by Hannah Arendt. New

York: Schocken, 1969. This much-cited essay on aesthetic theory argues that, with the modes of reproduction initiated by the Industrial Revolution, the reproduced work of art is incapable of retaining the "aura" of the original.

Web

http://www.artcyclopedia.com/artists/paik_nam_june.html. Artcyclopedia's entry on Paik, with links to several online exhibits, interviews, and fan pages.

http://www.dashsys.com/products/paik.html. Dash Systems Technical Support's page also has many links to online Paik resources. The company provides support for Paik's installations.

Audio/Visual

Videodrome. Directed by David Cronenberg. Starring James Woods, Debbie Harry. 87 minutes, rated R. Universal Pictures, 1983. DVD, videocassette. This sci-fi cult classic (about an insidious television broadcast that adversely affects the minds of its viewers) crosses the technology/body boundary as much of Paik's work does, but with a more horror-laden touch.

INTERNET
Art and Culture

ESSAY, 3 SCREEN SHOTS

Ellen Ullman's essay "The Museum of Me" single-handedly tackles a persistent argument about the Internet—should this new environment provide a space for genuine democratic community and communication, or should it be a frontier for profit and consumerism? Ullman most certainly sides with the first of these choices; as she sees it, the product-driven mind-set of those who are responsible for producing the technology has trickled down to the general public, creating an asocial society in which people sit in front of their screens, click up orders for goods and services, and forget what it is like to interact with others. Additionally, Ullman's concern is that this new model of capitalism, which does away with intermediary market forces such as the small-business owner, the factory worker, the distributor, and the retail salesclerk, is widening the gap between the haves and the have-nots of our society. The people partially responsible for fostering this economic shift, Ullman feels, are those already in privileged positions—people who can afford the fastest, most expensive technology, those who can afford to stay at home and click to shop. You might introduce this selection by having your students work in small groups to develop problem-solving presentations for Ullman's quandary—how can the World Wide Web become a truly democratic space?

After reading Ullman's essay, your students should visit the Art and Culture Web site (see URL below) and examine its design (and its apparent user-centered approach for wading through mounds of content). Ask them if Ullman might criticize this space on the Web or might wish for more like it. In discussing this topic, you can complicate the either/or stance by pointing out the vast array of reliable information about the visual and performing arts—the content this site offers is indeed quite impressive. Yet the fact that the site is full of commercial links reinforces Ullman's critique of a corporate-driven Internet, and the fact that the site uses memory-hungry Flash animation suggests that certain machines (and, it follows, certain people of privileged positions) are better able to access the site's data than others. Still, do these factors entirely invalidate this wonderful resource for cultural information?

MESSAGE

The advertising platitude of personal liberation fostered by Internet marketing still rings loudly today—"my" portals, customizable content sites, and commercials depicting individuals moving effortlessly through the hustle and bustle of the ordinary world thanks to Internet services that are still very much with us even now. Ullman's objection is that these messages are becoming less and less accurate as the Web becomes more and more commercialized. We are being saturated with messages of false freedom while the once-true freedoms of the Internet are quietly slipping away.

METHOD

Ullman's argument is that the phenomenon of a commercially sponsored cultural isolationism is a process that has unfolded over time, so her point is to show how our culture reached this point through incremental movements. A nonsequential arrangement of anecdotes might not fully convey this effect, but you could consider an in-class exercise wherein students reorder the anecdotes to bring about different effects.

MEDIUM

As facilitator, you should point out that, despite Ullman's criticisms, the Web contains a vast array of content. Try suggesting that your students consider couching their position as an argument of degree instead of an either/or stance.

ADDITIONAL WRITING TOPICS

1. In a short reflective piece, describe some of your best experiences using the World Wide Web or other Internet-based technologies. Describe some of your worst experiences as well.
2. In a comment accompanying this selection, Ellen Ullman expresses her disdain at what the Internet is becoming. She writes, "It's enormously sad to see the Internet being turned into the world-wide infomercial. The scariest part is the way Web site owners speak unabashedly about blurring the lines between editorial content and advertising, eagerly looking forward to the Web as a giant product-placement opportunity." Spend an hour or so exploring the Internet with Ullman's claim in mind. Are you able to find sites on the Web that do not fit her description? If so, give some examples of these sites, describing their design and content.
3. Ullman's essay ponders the value in the Internet's ability to allow for specific customization of content—the "museums of George and Mary and Helene." With what content (images, video, sound files, text) might your own "museum of me" be filled?

DIVERGENCES

Print

Ullman, Ellen. *Close to the Machine: Technophilia and Its Discontents*. San Francisco: City Lights Books, 1997.

Web

http://www.artandculture.com. The screen shots in the text are just a very few samples from the Art and Culture Web site.

http://www.cyberliteracy.net/. Laura Gurak's *Cyberliteracy: Navigating the Net with Awareness*, available online, offers useful terminology and concepts for describing the rhetorical construction of cyberculture.

http://www.ibiblio.org/stayfree/archives/15/ellen.html. This URL takes you directly to a brief interview with Ellen Ullman conducted by *Stay Free!*, an online journal on advertising culture.

Audio/Visual

eXistenZ. Directed by David Cronenberg. Starring Jennifer Jason Leigh, Jude Law. Rated R. Dimension Films, 1999. DVD. A computer programmer creates a virtual-reality video game that taps into the players' minds with violent consequences.

Divergences
Additional Resources for Reading Online, on Screen, on Paper

PRINT

Abbott, Edwin A. *Flatland: A Romance of Many Dimensions*. 1884. Reprint, New York: Dover Thrift Editions, 1992.

Adorno, Theodor. "Commitment." In *Art in Theory: 1900–1990*, edited by Charles Harrison and Paul Wood. Cambridge, England: Blackwell, 1992.

Alexie, Sherman. *Indian Killer*. New York: Atlantic Monthly Press, 1996.

———. *The Lone Ranger and Tonto Fistfight in Heaven*. New York: Atlantic Monthly Press, 1993.

———. *One Stick Song*. New York: Hanging Loose Press, 2000.

———. *The Toughest Indian in the World*. New York: Atlantic Monthly Press, 2000.

Allison, Dorothy. *Skin: Talking about Sex, Class, and Literature*. New York: Firebrand, 1994.

Alvarez, Julia. *How the Garcia Girls Lost Their Accents*. Chapel Hill, N.C.: Algonquin Books of Chapel Hill, 1991.

———. *In the Time of the Butterflies*. Chapel Hill, N.C.: Algonquin Books of Chapel Hill, 1994.

———. *Yo!* Chapel Hill, N.C.: Algonquin Books of Chapel Hill, 1997.

Anderton, Frances. *Las Vegas: The Success of Excess*. London: Ellipsis London, 1997.

Balmori, Diana, and Margaret Morton. *Transitory Gardens, Uprooted Lives*. New Haven: Yale University Press, 1993.

Barnard, Malcolm. *Art, Design, and Visual Culture*. New York: St. Martin's Press, 1998.

Baudrillard, Jean. *The Gulf War Did Not Take Place*. Translated by Paul Patton. Bloomington: Indiana University Press, 1995.

Benjamin, Walter. "The Work of Art in the Age of Mechanical Reproduction." In *Illuminations*, edited by Hannah Arendt. New York: Schocken, 1969.

Berger, John. *Ways of Seeing*. London: Penguin Group, 1972.

Bolter, Jay David, and Richard Grusin. *Remediation: Understanding New Media*. Cambridge, Mass.: MIT Press, 2000.

Breslin, Jimmy, and Stanley Crouch. "The Bad News: The Good News." *Esquire* (December 1995): 108 ff.

Brinkley, Douglas, et al. *36 Days: The Complete Chronicle of the 2000 Presidential Election Crisis*. New York: Times Books, 2001.

Brooks, Gwendolyn. *A Street in Bronzeville*. New York: Harper's, 1945.

Buchloh, Benjamin. "Interview with Gerhard Richter." In *Gerhard Richter: Paintings*, edited by I. Michael Danoff, Roald Nasgaard, and Terry Neff. London: Thames & Hudson, 1988.

Carnes, Mark C. "Past Imperfect: History According to the Movies." *Cineaste* 22, no. 4 (Fall 1996): 33–38.

Carnes, Mark C., ed. *Novel History: Historians and Novelists Confront America's Past (and Each Other)*. New York: Simon and Schuster, 2001.

Carnes, Mark C., ed. *Past Imperfect: History According to the Movies*. New York: Henry Holt, 1995.

Carruthers, Mary. *The Book of Memory: A Study of Memory in Medieval Culture*. Cambridge, England: Cambridge University Press, 1990.

Cofer, Judith Ortiz. *The Latin Deli*. New York: W. W. Norton, 1993.

Cohen, Herbert, ed. *125 Years of Famous Pages from* The New York Times: *1851–1976*. New York: Arno Press, 1976.

Conroy, Pat. *Beach Music*. New York: N. A. Talese, 1995.

———. *The Great Santini*. Boston: Houghton Mifflin, 1976.

———. *The Lords of Discipline*. Boston: Houghton Mifflin, 1980.

———. *My Losing Season*. Forthcoming.

———. *The Prince of Tides*. Boston: Houghton Mifflin, 1986.

———. *The Water Is Wide*. Boston: Houghton Mifflin, 1972.

Custer, George Armstrong. *My Life on the Plains—Or, Personal Experiences with Indians*. 1874. Reprint, Norman: University of Oklahoma Press, 1962.

Danticat, Edwidge. *The Farming of Bones*. New York: Abacus, 1999.

Doty, William. *Myths of Masculinity*. New York: Crossroad, 1993.

Eighner, Lars. "On Dumpster Diving." In *Homelessness: New England & Beyond*, edited by Padraig O'Malley. Amherst, Mass.: University of Massachusetts Press, 1992.

Ellis, Brett Easton. *American Psycho*. New York: Vantage, 1992.

Ellis, John. "Stars as a Cinematic Phenomenon." In *Film Theory and Criticism: Introductory Readings*, edited by Leo Braudy and Marshall Cohen. New York: Oxford University Press, 1999.

Ellison, Ralph. *The Collected Essays of Ralph Ellison*. New York: Modern Library, 1995.

———. *Flying Home and Other Stories*. Edited by John F. Callahan. New York: Random House, 1996.

———. *Invisible Man*. New York: Vintage International, 1995.

———. *Shadow and Act*. New York: Random House, 1964.

Epstein, Mitch. *The City*. New York: Powerhouse, 2001.

———. *Fire, Water, Wind*. Tenri-shi, Japan: Tenrikyo Doyusha, 1996.

———. *In Pursuit of India*. New York: Aperture, 1987.

———. *Vietnam: A Book of Change*. New York: W. W. Norton, 1996.

Faller, Kathleen C. *Child Sexual Abuse: An Interdisciplinary Manual for Diagnosis, Case Management, and Treatment*. New York: Columbia University Press, 1988.

———. *Child Sexual Abuse: Intervention and Treatment* Issues. Washington, D.C.: U.S. Department of Health and Human Services, 1993.

Faulkner, William. "The Bear." In *The Faulkner Reader: Selections from the Works of William Faulkner*. New York: Random House, 1954.

Frank, Anne. *The Diary of Anne Frank: The Critical Edition*. New York: Doubleday, 1989.

Frank, Thomas, and Matt Weiland, eds. *Commodify Your Dissent: Salvos from* The Baffler. New York: W. W. Norton, 1997.

Franken, Al. *Rush Limbaugh Is a Big Fat Idiot and Other Conversations*. New York: Delacorte Press, 1996.

Frazier, Ian. *Coyote vs. Acme*. New York: Farrar, Straus and Giroux, 1996.

———. *Dating Your Mom*. New York: Farrar, Straus and Giroux, 1986.

———. *Family*. New York: Farrar, Straus and Giroux, 1994.

———. *Great Plains*. New York: Farrar, Straus and Giroux, 1989.

———. *On the Rez*. New York: Farrar, Straus and Giroux, 2000.

Frost, Robert. *Complete Poems of Robert Frost*. New York: Holt, Rinehart and Winston, 1964.

Gamson, Joshua. *Freaks Talk Back: Tabloid Talk Shows and Sexual Nonconformity*. Chicago: University of Chicago Press, 1998.

Garcia, Leonardo. *Advocacy Advertising*. Manila: De La Salle University Press, 1999.

Gerber, Jurg, and Eric L. Jensen, eds. *Drug War, American Style: The Internationalization of Failed Policy and Its Alternatives*. New York: Garland, 2001.

Goodrich, Norma L. *Heroines: Demigoddess, Prima Donna, Movie Star*. New York: HarperCollins, 1993.

Greenberg, Clement. "Avant-Garde and Kitsch." *Partisan Review* 1, no. 5 (Fall 1939): 34–49.

Heller, Steven. *Design Literacy (Continued): Understanding Graphic Design*. New York: Allworth Press, 1999.

Hilger, Michael. *From Savage to Nobleman: Images of Native Americans in Film*. Metuchen, N.J.: Scarecrow Press, 1995.

Hohm, Charles F., and Lori J. Jones, eds. *Population: Opposing Viewpoints*. San Diego: Greenhaven Press, 1995.

hooks, bell. *About Love: New Visions*. New York: William Morrow, 2000.

———. *Ain't I a Woman?: Black Women and Feminism*. Boston: South End Press, 1981.

———. *Bone Black: Memories of Girlhood*. New York: Henry Holt, 1996.

———. *Feminist Theory: From Margin to Center*. Boston: South End Press, 1984.

———. *Killing Rage: Ending Racism*. New York: Henry Holt, 1995.

———. *Outlaw Culture: Resisting Representations*. New York: Routledge, 1994.

———. *Talking Back: Thinking Feminist, Thinking Black*. Boston: South End Press, 1988.

———. *where we stand: class matters*. New York: Routledge, 2000.

———. *Wounds of Passion: A Writing Life*. New York: Henry Holt, 1997.

———. *Yearning: Race, Gender, and Cultural Politics*. Boston: South End Press, 1990.

Iyer, Pico. *Falling Off the Map*. New York: Random House, 1993.

———. *The Global Soul: Jet Lag, Shopping Malls, and the Search for Home*. New York: Random House, 2000.

———. *The Lady and the Monk: Four Seasons in Kyoto*. New York: Random House, 1991.

Kafka, Franz. *Amerika*. Translated by Willa and Edmund Muir. New York: Schocken Books, 1996.

———. *The Castle*. Translated by Mark Harman. New York: Schocken Books, 1998.

———. *The Metamorphosis, In the Penal Colony, and Other Stories: With Two New Stories*. Translated by Joachim Neugroschel. New York: Scribner, 2000.

———. *The Trial*. Translated by Willa and Edmund Muir. New York: Knopf, 1972.

Kruger, Barbara. *Thinking of You*. Cambridge, Mass.: MIT Press, 1999.

Kunz, Martin, ed. *William Wegman : Paintings, Drawings, Photographs, Videotapes*. New York: Abrams, 1990.

Lahiri, Jhumpa. *Interpreter of Maladies*. New York: Meridian, 2000.

Lasn, Kalle. *Culture Jam: The Uncooling of America*. New York: HarperCollins, 2001.

Lawlor, Laurie. *Where Will This Shoe Take You?: A Walk through the History of Footwear*. New York: Walker & Co., 1996.

Levin, Gail, ed. *Silent Places: A Tribute to Edward Hopper*. New York: Universe, 2000.

Limbaugh, Rush. *The Way Things Ought to Be*. New York: Pocket Books, 1992.

Lin, Maya. *Boundaries*. New York: Simon and Schuster, 2000.

Lopez, Erika. *Flaming Iguanas: An Illustrated All-Girl Road Novel Thing*. New York: Scribners, 1998.

Lyon, Danny. *The Bikeriders*. New York: Macmillan, 1968.

———. *Conversations with the Dead*. New York: Holt, Rinehart and Winston, 1971.

———. *The Destruction of Lower Manhattan*. New York: Macmillan, 1969.

———. *I Like to Eat Right on the Dirt: A Child's Journey Back in Space and Time*. New York.: Bleak Beauty: Filmhaus, 1989.

———. *Knave of Hearts*. Santa Fe, N.M.: Twin Palms Press, 1999.

———. *Pictures from the New World*. New York: Aperture, 1981.

Mann, Sally. *At Twelve: Portraits of Young Women*. New York: Aperture, 1988.

———. *Immediate Family*. New York: Aperture, 1992.

Manson, Marilyn. *The Long Hard Road Out of Hell*. New York: Reagan Books, 1999.

Mark, Mary Ellen. *American Odyssey*. New York: Aperture, 1999.

———. *Streetwise*. New York: Aperture, 1988.

McCloud, Scott. *Understanding Comics: The Invisible Art*. New York: HarperCollins, 1993.

Meisel, Louis K. *Photo-Realism*. New York: Abradale, 1989.

Merkl, Peter H. *A Coup Attempt in Washington?: A European Mirror on the 1998–1999 Constitutional Crisis*. New York: Palgrave, 2001.

Miller, Wayne. *Chicago's South Side: 1946–1948*. Berkeley: University of California Press, 2000.

Morgan, Stuart, ed. *Damien Hirst: No Sense of Absolute Corruption*. London: Gagosian Gallery, 1996.

Morton, Margaret. *Fragile Dwelling*. New York: Aperture, 2000.

———. *The Tunnel: The Underground Homeless of New York City*. New Haven: Yale University Press, 1995.

Mulvey, Laura. "Visual Pleasure and Narrative Cinema." *Screen* 16, no. 3 (1975): 6–18. Reprinted in Laura Mulvey, *Visual and Other Pleasures*. Bloomington: Indiana University Press, 1989.

Oates, Joyce Carol. *Blonde*. New York: Echo, 2000.

———. *Broke Heart Blues: A Novel*. New York: Dutton, 1999.

———. *Them*. New York: Vanguard, 1969.

O'Hara, Frank. *The Collected Poems of Frank O'Hara*. Edited by Donald Allen. New York: Knopf, 1971.

Orwell, George. *1984*. 1949. Reprint, London: Secker & Warburg, 1987.

Patterson, Vivian. *Carrie Mae Weems: The Hampton Project*. New York: Aperture, 2001.

Richter, Gerhard. *The Daily Practice of Painting: Writings 1962–1993*. Edited by Hans-Ulrich Obrist. Cambridge, Mass.: MIT Press, 1995.

Riis, Jacob. *How the Other Half Lives: Studies among the Tenements of New York*. New York: Dover, 1971.

Rodriguez, Joseph. *East Side Stories: Gang Life in East L.A.* New York: Powerhouse Books, 1996.

———. *Spanish Harlem*. Washington, D.C.: National Museum of American Art, 1994.

Rodriguez, Richard. *Days of Obligation: An Argument with My Mexican Father*. New York: Viking, 1992.

———. *The Hunger of Memory: The Education of Richard Rodriguez*. Boston: D. R. Godine, 1982.

Roiphe, Katie. *Last Night in Paradise: Sex and Morals at the Century's End*. Boston: Little, Brown, 1997.

———. *The Morning After: Sex, Fear, and Feminism on Campus*. Boston: Little, Brown, 1993.

Sante, Luc. *Evidence*. New York: Farrar, Straus and Giroux, 1992.

———. *Factory of Facts*. New York: Pantheon, 1998.

———. *Low Life: Lures and Snares of Old New York*. New York: Farrar, Straus and Giroux, 1991.

Schlesinger, Arthur. *The Age of Jackson*. New York: Book Find Club, 1945.

———. *The Disuniting of America*. New York: Norton, 1992.

———. *A Thousand Days: John F. Kennedy in the White House*. Boston: Houghton Mifflin, 1965.

Sedaris, David. *Barrel Fever: Stories and Essays*. New York: Little, Brown, 1995.

———. *Holidays on Ice*. New York: Little, Brown, 1998.

———. *Me Talk Pretty One Day*. New York: Little, Brown, 2000.

Sexton, Anne. *The Awful Rowing toward God*. Boston: Houghton Mifflin, 1975; London: Chatto and Windus, 1977.

———. *The Book of Folly*. Boston: Houghton Mifflin, 1972; London: Chatto and Windus, 1974.

———. *The Complete Poems*. Boston: Houghton Mifflin, 1981.

———. *The Death Notebooks*. Boston: Houghton Mifflin, 1974; London: Chatto and Windus, 1975.

———. *45 Mercy Street*. Edited by Linda Gray Sexton. Boston: Houghton Mifflin, 1976; London: Martin Secker and Warburg, 1977.

———. *No Evil Star: Selected Essays, Interviews, and Prose*. Edited by Stephen E. Colburn. Ann Arbor: University of Michigan Press, 1985.

———. *Words for Dr. Y.: Uncollected Poems with Three Stories*. Edited by Linda Gray Sexton. Boston: Houghton Mifflin, 1978.

Sills, Leslie. *In Real Life: Six Women Photographers*. New York: Holiday House, 2000.

Spiegelman, Art. *Maus: A Survivor's Tale*. New York: Pantheon, 1991.

Steinem, Gloria. *Outrageous Acts and Everyday Rebellions*. New York: Holt, Rinehart, and Winston, 1982.

———. *Revolution from Within: A Book of Self-Esteem*. Boston: Little, Brown, 1992.

Sullivan, Andrew. *Virtually Normal: An Argument about Homosexuality and Love Undetectable*. New York: Knopf, 1995.

Teitelbaum, Matthew, ed. *Montage and Modern Life, 1919–1942*. Cambridge, Mass.: MIT Press, 1992.

Thoreau, Henry David. Walden *and* Civil Disobedience*: Complete Texts with Introduction, Historical Contexts, Critical Essays*. Edited by Paul Lauter. Boston: Houghton Mifflin, 2000.

Toscani, Oliviero. *Current Biography*. (Sept. 1998): 55ff.

Trudeau, Garry. *Duke 2000*. Los Angeles: Andrews McMeel, 2000.

Twain, Mark. *The Adventures of Huckleberry Finn*. 1884. Reprint, New York: Dale Books, 1978.

Twitchell, James B. *Adcult USA: The Triumph of Advertising in American Culture*. New York: Columbia University Press, 1996.

———. *Twenty Ads That Shook the World: The Century's Most Groundbreaking Advertising and How It Changed Us All*. New York: Crown, 2000.

Ullman, Ellen. *Close to the Machine: Technophilia and Its Discontents*. San Francisco: City Lights Books, 1997.

Updike, John. *Rabbit Angstrom: A Tetralogy*. New York: Random House, 1995.

Walker, John, and Sarah Chaplin. *Visual Culture: An Introduction*. New York: St. Martin's Press, 1997.

Waplington, Nick. *The Indecisive Memento*. New York: Aperture, 1999.

———. *Living Room*. New York: Aperture, 1991.

———. *Other Edens*. New York: Aperture, 1994.

———. *The Wedding: New Pictures from the Continuing "Living Room" Series*. New York: Aperture, 1996.

Warhol, Andy. *Shoes, Shoes, Shoes*. New York: Bulfinch Press, 1997.

Weegee. *Naked City*. New York: Essential Books, 1945.

Wegman, William. *Cinderella*. New York: H. N. Abrams, 1993.

———. *Fashion Photographs*. New York: H. N. Abrams, 1999.

———. *Little Red Riding Hood*. New York: H. N. Abrams, 1993.

———. *Man's Best Friend*. New York: H. N. Abrams, 1982.

Weiss, Jessica. *To Have and to Hold: Marriage, the Baby Boom, and Social Change*. Chicago: University of Chicago Press, 2000.

Welch, Kathleen. *Electric Rhetoric: Classical Rhetoric, Oralism, and a New Literacy*. Cambridge, Mass.: MIT Press, 1999.

Wicks, Robert. *Understanding Audiences: Learning to Use the Media Constructively*. Mahwah, N.J.: Lawrence Erlbaum, 2001.

Wiesel, Elie. *Night*. New York: Bantam Books, 1982.

Williams, Terry Tempest. *Coyote's Canyon*. New York: Gibbs Smith, 1999.

———. *Desert Quartet*. New York: Pantheon, 1995.

———. *Leap*. New York: Pantheon, 2000.

———. *Pieces of a White Shell: A Journey to Navajoland*. New York: Scribner, 1984.

———. *Refuge: An Unnatural History of Family and Place*. New York: Pantheon, 1991.

———. *An Unspoken Hunger: Stories from the Field*. New York: Pantheon, 1994.

Zimbardo, Xavier. *India Holy Song*. New York: St. Martin's Press, 2000.

WEB

http://abcnews.go.com/sections/world/DailyNews/germany010329_body.html. ABC News.com article on "Body Worlds" art exhibit.

http://dairyland.com. Web site for Dairyland Blog.

http://db.education-world.com/perl/browse?cat_id=1135. The Education World Web site.

http://dive.woodstock.edu/~dcox/ohenry/cofer.html. Montgomery College's Web page on Judith Ortiz Cofer.

http://docs.yahoo.com/info/misc/history.html. Yahoo!'s media relations page.

http://eclipse.barnard.columbia.edu/~as833/bc3401/. A hypertextual essay by a Columbia University student on gang culture.

http://faculty.washington.edu/rods/art_lyon.html. Rod Slemmons's essay on Danny Lyon.

http://hardpress.com/newhp/lingo/authors/morton.html. Hardpress's online photo-essay by Margaret Morton entitled "José Camacho's House."

http://home.pacifier.com/~paddockt/sedaris.html. The "Unofficial David Sedaris Internet Resource."

http://home.sprynet.com/~mersault/sexton/. An Anne Sexton bibliography.

http://members.tripod.com/mking60/conspiracies.htm. A personal Web site compiling several conspiracy theories and global threats, the "atomic secrets" Luc Sante's essay mentions.

http://owa.chef-ingredients.com/postUK/20/harvey.htm. Chef Ingredients page on Ellen Harvey.

http://parallel.park.uga.edu/~jcofer/. Judith Ortiz Cofer's University of Georgia faculty page.

http://slate.msn.com/. *Slate*, the Microsoft-sponsored online politics and culture magazine.

http://storm.usfca.edu/~southerr/jco.html. University of San Francisco's Celestial Timepieces' Joyce Carol Oates page.

http://thewall-usa.com. The official site for the Vietnam Veterans Memorial.

http://voices.cla.umn.edu/authors/bellhooks.html. University of Minnesota's "Voices from the Gap" site.

http://vos.ucsb.edu/shuttle/media.html. "Voice of the Shuttle: Web Page for Humanities Resources."

http://wings.buffalo.edu/cas/english/faculty/conte/syllabi/377/Frank_O'Hara.html. A University of Buffalo site with Frank O'Hara's poem.

http://www.2600.com. The online hacker magazine *2600*.

http://www.4threvolution.com. 4th Revolution's official portfolio site.

http://www.aap.org/advocacy/mediamatters.htm. The American Academy of Pediatrics' National Media Education Campaign.

http://www.about.com. About.com search engine portal.

http://www.aclu.org/. The American Civil Liberties Union home page.

http://www.adbusters.org/. The Adbusters' home page.

http://www.adobe.com. The home page for Adobe Systems Incorporated.

http://www.amtrak.com. Amtrak's home page.

http://www.annegeddes.com/. Anne Geddes's official site.

http://www.aperture.org/. The home page for *Aperture* magazine.

http://www.apple.com/applemasters/dhirst/. Apple's "Think Different" online entry on Damien Hirst.

http://www.ardemgaz.com/prev/clinton/aapaula14sidea.html. Associated Press excerpt from President Bill Clinton's deposition in the Paula Jones case.

http://www.artandculture.com. The Art and Culture Web site.

http://www.artchive.com/artchive/J/johnsbio.html. The Artchive site's entry on Jasper Johns.

http://www.artchive.com/artchive/R/richter.html. The Artchive Web site's entry on Gerhard Richter.

http://www.artcyclopedia.com/artists/estes_richard.html. Artcyclopedia's entry on Richard Estes.

http://www.artcyclopedia.com/artists/hopper_edward.html. Artcyclopedia's entry on Edward Hopper.

http://www.artcyclopedia.com/artists/koons_jeff.html. Artcyclopedia's entry on Jeff Koons.

http://www.artcyclopedia.com/artists/paik_nam_june.html. Artcyclopedia's entry on Nam June Paik.

http://www.arts.monash.edu.au/visarts/globe/issue4/bkrutit.html. Images from a 1996 Barbara Kruger installation in Melbourne.

http://www.artshum.org/pages/dream.html. Jan Kurtz's online essay "Dream Girls: Women in Advertising."

http://www.banmines.org. PALM's official site.

http://www.benetton.com. The official United Colors of Benetton Web site.

http://www.benetton.com/colors. UCB's online magazine *Colors*.

http://www.blogger.com. Blogger.com home page.

http://www.bookpage.com/0006bp/david_sedaris.html. Bookpage's interview with David Sedaris.

http://www.bookwire.com/bbr/bbr-home.html. *Boston Book Review* home page.

http://www.britneyspears.com. The official Britney Spears Web site.

http://www.care.org. This is the site for global relief organization Cooperative for Assistance and Relief Everywhere (CARE).

http://www.chicagohs.org/. Web site for the Chicago Historical Society.

http://www.chickenhead.com/truth/1950s.html. The "Truth in Advertising" site.

http://www.civitas.org/. The CIVITAS Web site.

http://www.collegefund.org/main.shtm. Home page for the American Indian College Fund.

http://www.com.washington.edu/rccs/. The Resource Center for Cyberculture Studies (RCCS).

http://www.coolmemes.com/reader/conroy.htm. The Cool Memes page of Pat Conroy–related links.

http://www.coolmemes.com/reader/sexton.htm. The Cool Memes page on Anne Sexton.

http://www.cooper.edu/art/nyphoto/. Cooper Union School of Art's "New York Photographs" exhibit.

http://www.crosswinds.net/~russ_posters/. An online gallery of Soviet, Czech, Polish, and Cuban communist propaganda posters.

http://www.cs.technion.ac.il/~eckel/Kafka/kafka.hml. Site with Kafka images.

http://www.cyberliteracy.net/. Laura Gurak's *Cyberliteracy: Navigating the Net with Awareness*.

http://www.dashsys.com/products/paik.html. Dash Systems Technical Support's home page.

http://www.dazereader.com/sallymann.htm. Daze Reader's online resources on Sally Mann.

http://www.depauwgallery.com/Top/artists/artist12.html. Depauw Gallery's exhibit of Nick Waplington's work.

http://www.diacenter.org/exhibs/richter/richter.html. New York's Dia Center for the Arts' Richter exhibit.

http://www.dieter-obrecht.com/richter/richter.htm. Wildbrush's art.today, a Gerhard Richter fan site.

http://www.doonesbury.com/. The Doonesbury Electronic Town Hall.

http://www.doubletakemagazine.org/. *Doubletake* magazine's Web site.

http://www.eng.fju.edu.tw/Literary_Criticism/feminism/index.html. Fu Jen Catholic University's site on feminism and gender studies.

http://www.english.uiuc.edu/maps/poets/m_r/ohara/ohara.htm. Modern American Poetry's entry on Frank O'Hara.

http://www.english.upenn.edu/~afilreis/50s/ellison-main.html. The University of Pennsylvania's critical commentary site on *Invisible Man*.

http://www.excite.com. Excite search engine portal.

http://www.freedomship.com/. The Freedom Ship Project home page.

http://www.gangsorus.com/gangs/ganggirls.html. The "Gangs or Us" Web site.

http://www.geocities.com/SoHo/Cafe/9747/kruger.html. A fan site with a short personal essay on Barbara Kruger's art.

http://www.goarmy.com. The U.S. Army's recruitment Web site.

http://www.greatdreams.com/princess.htm. A fan's site about Princess Diana.

http://www.hewlettpackard.com/. Hewlett-Packard's home page.

http://www.historyinfilm.com/. The History in Film site.

http://www.ibiblio.org/stayfree/archives/15/ellen.html. A brief interview with Ellen Ullman by *Stayfree!*

http://www.icp.org/weegee/. The International Center of Photography's Weegee site.
http://www.ihrinfo.ac.uk/maps/. "The Map History/History of Cartography" home page.
http://www.internationalposter.com. International Poster home page.
http://www.jennicam.com. The JenniCam site.
http://www.jokeindex.com/joke.asp?Joke = 2795. Jokeindex.com's parody list of the thirty-one worst Norman Rockwell paintings.
http://www.kodak.com/US/en/corp/aboutKodak/kodakHistory/kodakHistory.shtml. Kodak's history of its technological, cultural, and corporate milestones.
http://www.levity.com/corduroy/sexton.htm. Anne Sexton fan site.
http://www.lib.berkeley.edu/EART/aerial.html. University of California, Berkeley's Aerial Photography and Satellite Imagery site.
http://www.loc.gov/exhibits/treasures/trm003p.html. The Library of Congress site on the development of the *Vietnam Veterans Memorial*.
http://www.lucent.com. Lucent Technologies' home page.
www.marilynmanson.net/. Official Marilyn Manson site.
http://www.maryellenmark.com/. Mary Ellen Mark's home page.
http://www.motherjones.com. The site for *Mother Jones*.
http://www.moviejuice.com/elian/moviereport.htm. A spoof site featuring movie reviews by "Elián Gonzáles."
http://www.msmagazine.com/. *Ms.* magazine's online site.
http://www.msnbc.com/news/NW-front_ Front.asp. The *Newsweek* and MSNBC partner site.
http://www.nara.gov/exhall/picturing_the_century/portfolios/port_lyon.html. National Archives' "Picturing the Century," with a portfolio of Danny Lyon's work.
http://www.nerve.com. Nerve.com's provocative site.
http://www.newspeakdictionary.com/. The Newspeak Dictionary site.
http://www.nikeid.com. The Nike iD site.
http://www.npg.org. The home page of Negative Population Growth.
http://www.npr.com/. The NPR Online home page.
http://www.npr.org/programs/death/readings/stories/sante.html. NPR Online's transcript and audio file of Luc Sante's "The Unknown Soldier."
http://www.nrm.org/. The Norman Rockwell Museum at Stockbridge online.
http://www.ocaiw.com/mapple.htm. Site with online resources on Robert Mapplethorpe.
http://www.outwestnewspaper.com/airstream.html. *Outwest News*'s page on the Airstream Trailer.
http://www.paonline.com/zaikoski/rockwell.htm. A site with several high-resolution images of *Post* covers.
http://www.pathtofreedom.com/main.html. Path to Freedom home page.
http://www.pbs.org/conjure/cm.html. PBS's online "Conjure Women" series profile of Carrie Weems.
http://www.pbs.org/newshour/character/links/nixon_speech.html. PBS Online's transcript of Richard Nixon's 1974 resignation speech.
http://www.peta.org/. Home page for the People for the Ethical Treatment of Animals.
http://www.pitt.edu/~kafka/intro.html. University of Pittsburg site dedicated to Kafka studies.
http://www.ps1.org/cut/studioimg/harvey.html. The International Studio Program's site, which includes a detailed résumé of Ellen Harvey.
http://www.salon.com/books/int/1998/03/cov_si_31ntb.html. Salon.com interview with Dorothy Allison.
http://www.salon.com/directory/topics/david_sedaris/. Salon.com's list of David Sedaris texts and audio files.
http://www.salon.com/people/bc/2000/03/28/mark/. Salon.com's profile on Mary Ellen Mark.
http://www.sandcreek.org/. The home page for the Northern Cheyenne Sand Creek Massacre Site Project.
http://www.shey.net/niked.html.Original e-mail correspondence between Jonah Peretti and Nike.

http://www.sscnet.ucla.edu/southasia/Culture/Cinema/cinema.html. The Manas site on Indian film and culture.

http://www.texaschainsawmassacre.net. Fan site for film *Texas Chainsaw Massacre*.

http://www.theonion.com. Online version of *The Onion*.

http://www.thetruth.com. The Truth home page.

http://www.thevirtualwall.org. A virtual tour of the *Vietnam Veterans Memorial*.

http://www.thirdworldtraveler.com/Blum/DominicanRepublic_KH.html. A selection from William Blum's *Killing Hope*.

http://www.time.com. *Time* Online.

http://www.uiowa.edu/~commstud/adclass/1984_mac_ad.html. Sarah Stein's essay on Apple's "1984" ad.

http://www.unitedmedia.com/comics/. United Media's Comics.com site.

http://www.universes-in-universe.de/america/us_afro/weems/weems1.htm. Interview with Carrie Weems conducted by Kaira M. Cabañas of Universes in Universe.

http://www.ushmm.org/museum. The United States Holocaust Memorial Museum Web site.

http://www.uta.edu/english/V/test/agamben/v.1.html. Rhetoric scholar Victor Vitanza's web presentation on "Objects and Whatever-Beings: The Coming (Educational) Community."

http://www.washingtonpost.com/wp-srv/national/longterm/meltingpot/melt0222.htm. The *Washington Post*'s series of articles on the "Myth of the Melting Pot."

http://www.webbys.com. Web site for the Webby Awards.

http://www.wegmanworld.com/. William Wegman's official site.

http://www.wellsfargo.com. Wells Fargo home page.

http://www.whatsyourantidrug.com. The Freevibe Web site.

http://www.zpg.org. Home page of Zero Population Growth.

AUDIO/VISUAL

The Abandoned Children. Directed by Danny Lyon. 63 minutes, not rated. 1974. Videocassette.

Antichrist Superstar. Music and lyrics by Marilyn Manson. Interscope Records, 1996. Compact disc.

Barbara Kruger: Pictures and Words. 28 minutes. Facets Multimedia, 1996. Available from http://www.buyindies.com. Videocassette.

Bastard Out of Carolina. Directed by Angelica Huston. 101 minutes, rated R. Fox Lorber, 1996. Videocassette.

Behind the Scenes with Carrie Mae Weems. Directed by Ellen Hovde and Muffie Meyer. 30 minutes, not rated.

bell hooks on Video: Cultural Criticism & Transformation. 70 minutes, not rated. Facets Multimedia. Available from http://www.buyindies.com.

Beyond Killing Us Softly: The Strength to Resist. Directed by Margaret Lazarus. Not rated. Videocassette.

Blade Runner. Directed by Ridley Scott. Starring Harrison Ford. 117 minutes, rated R. Warner Home Video, 1991 (filmed in 1982). Videocassette.

Blonde. Starring Patrick Dempsey. 4 hours, not rated. CBS, 2001.

Boys Don't Cry. Directed by Kimberly Pierce. Starring Hilary Swank, Chloe Sevigny. 116 minutes, rated R. Twentieth Century Fox, 1999. Videocassette, DVD.

Boyz N the Hood. Directed by John Singleton. Starring Ice Cube, Lawrence Fishburne, Cuba Gooding Jr. 112 minutes, rated R. Columbia/Tristar, 1992. Videocassette.

California vs. O. J. Simpson, vols. 1 and 2. 465 minutes, not rated. MPI Home Videos, 1995. Videocassette.

Casualties of War. Directed by Brian De Palma. Starring Sean Penn, Michael J. Fox. 120 minutes, rated R. Columbia Pictures, 1989. Videocassette.

Celebrity. Directed by Woody Allen. Starring Leonardo DiCaprio, Hank Azaria, Winona Ryder. 113 minutes, rated R. Mirimax, 1997. Videocassette, DVD.

Chutney Popcorn. Directed by Nisha Ganatra. Starring Ganatra, Jill Hennessy, Madhur Jaffrey, Sakina Jaffrey. Rated PG. Mata Productions, Inc., 1999. Videocassette.

A Clockwork Orange. Directed by Stanley Kubrick. 137 minutes, rated R. Warner Home Video, 1991 (filmed in 1971). Videocassette.

Crazy People. Directed by Tony Bill. Starring Dudley Moore. 90 minutes, rated PG. Paramount, 1990. Videocassette.

Crumb. Directed by Terry Zwigoff. Rated R. Columbia/Tristar, 1995. Videocassette.

The David Sedaris Box Set. Time Warner Audio Books, 2000.

Daydream Nation. Music and lyrics by Sonic Youth. Blast First/Enigma Records, 1988. Compact disc.

Dead Man Walking. Directed by Tim Robbins. Starring Sean Penn, Susan Sarandon. 122 minutes, rated R. MGM/UA Studios, 1995. Videocassette.

Desert Blue. Directed by Morgan J. Freeman. Starring Brandon Sexton, Kate Hudson, Christina Ricci. DVD. 90 minutes, rated R. Columbia Tristar, 1999.

"Disappearing Pay Phones." National Public Radio, February 14, 2001. Audio recording.

El Otro Lado (The Other Side). Directed by Danny Lyon. 60 minutes, not rated. 1978. Videocassette.

eXistenZ. Directed by David Cronenberg. Starring Jennifer Jason Leigh, Jude Law. Rated R. Dimension Films, 1999. DVD.

Fight Club. Directed by David Fincher. Starring Brad Pitt, Edward Norton, Meat Loaf. 139 minutes, rated R. 20th Century Fox, 1999. DVD.

Girl, Interrupted. Directed by James Mangold. Starring Winona Ryder, Whoopi Goldberg, Angelina Jolie. 127 minutes, rated R. Columbia/Tristar, 2000. Videocassette.

Hackers. Directed by Iain Softley. Starring Angelina Jolie, Johnny Lee Miller. 104 minutes, rated PG-13. MGM/UA Studios, 1995. Videocassette.

Human Growth. 20 minutes. Journal Films, 1994. Videocassette.

The Human Sexes: A Natural History of Man and Woman, six volumes. Written and presented by Desmond Morris. Not rated. Discovery Communication, 1997. Videocassette.

Illusions. Directed by Julie Dash. 34 minutes, not rated. 1983.

Incident at Oglala: The Leonard Peltier Story. Directed by Michael Apted. Narrated by Robert Redford. 90 minutes, rated PG. Artisan, 1994. Videocassette.

Interviewing for Child Sexual Abuse: A Forensic Guide. Produced and directed by Kevin Dawkins. 35 minutes, not rated. Guilford Publications, 1998. Videocassette.

Interview with Christopher Darden on National Public Radio's *Fresh Air*. WHYY, Philadelphia, March 31, 1996. Audio recording.

Interview with Dorothy Allison on National Public Radio's *Fresh Air*. WHYY, Philadelphia, March 23, 1998. Audio recording.

Interview with Marilyn Manson on National Public Radio's *Fresh Air*. 60 minutes. WHYY, Philadelphia, February 24, 1998.

Interview with Sherman Alexie on National Public Radio's *Fresh Air*. WHYY, Philadelphia, September 21, 1993. Audio recording.

Jazz. Directed by Ken Burns. 19+ hours, not rated. PBS, 2001. Videocassette, DVD.

JFK. Directed by Oliver Stone. Starring Kevin Costner. 189 minutes, rated R. Warner Bros., 1992. Videocassette.

Kids. Directed by Larry Clark. 91 minutes, rated R. Trimark Studios, 1995. Videocassette.

The Killing Fields. Directed by Roland Joffé. Starring Sam Waterson, Haing S. Ngor. 142 minutes, rated R. Warner Studios, 1984. Videocassette.

Little Boy. Directed by Danny Lyon. 52 minutes, not rated. 1976. Videocassette.

Maya Lin: A Strong, Clear Vision. Directed by Frieda Lee Mock. Not rated. Ocean Releasing, 1994. Videocassette.

Menace II Society. Directed by Allen and Albert Hughes. Starring Larenz Tate, Tyrin Turner. 104 minutes, rated R. New Line Studios, 1993. Videocassette, DVD.

No Safe Place: Violence against Women. Produced by KUED, Salt Lake City, 1998. Videocassette.

Office Killer. Directed by Cindy Sherman. 83 minutes, rated R. Dimension, 1997. Videocassette.

Paris Is Burning. Directed by Jennie Livingston. 78 minutes, rated R. Fox/Lorber, 1991. Videocassette.

The Patriot. Directed by Roland Emmerich. Starring Mel Gibson. 165 minutes, rated R. Warner, 2000. DVD.

A Plague of Plastic Soldiers: Land Mines in Cambodia. Produced by Stephen Smith. Soundprint Media Center (1-888-38-TAPES), 1997. Audio recording.

Portrait of an Artist at Work: Cindy Sherman. Directed by Michel Auder. 1988. Videocassette.

A Raisin in the Sun. Directed by Daniel Petrie. Starring Sidney Poitier. 128 minutes. Columbia Pictures Corporation, 1961. Videocassette.

The Red Violin. Directed by François Girard. Starring Samuel Jackson. 143 minutes, rated R. Rhombus Media, 1998. Videocassette.

The Ringmaster. Directed by Neil Abramson. Starring Jerry Springer. 95 minutes, rated R. Artisan, 1998. Videocassette.

Ronald Reagan: The Great Speeches, vol. 1. 75 minutes. Speechworks. Audio recording.

"Salt Lake City Remote: The Changing West" on National Public Radio's *Talk of the Nation*. National Public Radio (http://www.npr.org), May 14, 1998. Audio recording.

Schindler's List. Directed by Stephen Spielberg. Starring Liam Neesom, Ralph Fiennes. 197 minutes, rated R. Universal, 1993. Videocassette.

The Simpsons. "Mom and Pop Art." 23 minutes. FOX Broadcasting, April 11, 1999.

Smoke Signals. Directed by Chris Eyre. 89 minutes, rated PG-13. Mirimax, 1998. Videocassette.

Texas Chainsaw Massacre. Directed by Tobe Hooper. 83 minutes, rated R. MPI Video, 1993. Videocassette.

Thrashin'. Directed by Peter Weathers. Starring Josh Brolin. Rated PG. 1986. Videocassette.

Titanic. Directed by James Cameron. Starring Leonardo DiCaprio, Kate Winslet. 192 minutes, rated PG-13. Paramount, 1998. Videocassette.

To Die For. Directed by Gus Van Sant. Starring Nicole Kidman. 105 minutes, rated R. Columbia, 1995. Videocassette.

Traffic. Directed by Stephen Soderbergh. Starring Catherine Zeta Jones, Michael Douglas, Benicio del Toro. 147 minutes, rated R. USA Films, 2001. DVD.

Tron. Directed by Steven Lisberger. Starring Jeff Bridges. 96 minutes, rated PG. Disney, 1982. Videocassette.

The Truman Show. Directed by Peter Weir. Starring Jim Carrey. 103 minutes, rated PG. Paramount, 1998. Videocassette.

Videodrome. Directed by David Cronenberg. Starring James Woods, Debbie Harry. 87 minutes, rated R. Universal Pictures, 1983. Videocassette, DVD.

Voices and Visions: Sylvia Plath. 60 minutes. Winstar Home Entertainment, 1999. Videocassette.

Wanderers. Directed by Philip Kaufman. Starring Ken Wahl. Rated R. Warner, 1992. Videocassette.

West Side Story. Directed by Robert Wise and Jerome Robbins. Starring Natalie Wood. 151 minutes, not rated. MGM/UA, 1961. Videocassette, DVD.